FEEDING THE CITY

FEEDING THE CITY:
WORK AND FOOD CULTURE OF
THE MUMBAI *DABBAWALAS*

Sara Roncaglia

OpenBook
Publishers

As with all Open Book Publishers titles, digital material and resources associated with this volume are available from our website at:
http://www.openbookpublishers.com/isbn/9781909254008

ISBN Hardback: 978-1-909254-01-5
ISBN Paperback: 978-1-909254-00-8
ISBN Digital (PDF): 978-1-909254-02-2
ISBN Digital ebook (epub): 978-1-909254-03-9
ISBN Digital ebook (mobi): 978-1-909254-04-6

DOI: 10.11647/OBP.0031

Cover image: Preparation of a meal in Mumbai, May 2007. Photo by Sara Roncaglia.

Translated from the Italian by Angela Arnone.

Typesetting by www.bookgenie.in

All paper used by Open Book Publishers is SFI (Sustainable Forestry Initiative), and PEFC (Programme for the Endorsement of Forest Certification Schemes) Certified.

Printed in the United Kingdom and United States by Lightning Source for Open Book Publishers (Cambridge, UK).

Contents

One blue-bright Bombay morning, in the middle of the masses on the street, I have a vision: that all these individuals, each with his or her own favourite song and hairstyle, each tormented by an exclusive demon, form but the discrete cells of one gigantic organism, one vast but singular intelligence, one sensibility, one consciousness. Each person is the end product of an exquisitely refined specialization and has a particular task to perform, no less and no more important than that of any other of the six billion components of the organism. It is a terrifying image; it makes me feel crushed, it eliminates my sense of myself, but it is ultimately comforting because it is such a lovely vision of belonging. All these ill-assorted people walking towards the giant clock on Churchgate: they are me; they are my body and my flesh. The crowd is the self, fourteen million avatars of it, fourteen million celebrations. I will not merge into them; I have elaborated myself into them. And if I understand them well, they will all merge back into me, and the crowd will become the self, one, many-splendoured.

Suketu Mehta, *Maximum City: Bombay Lost and Found* (London: Headline Review, 2005), p. 590.

Acknowledgements

This book is the product of a dual research effort: its subject matter was the topic of my PhD in Political, Social and Psychological Sciences at Genoa University under the supervision of Marco Aime, Professor in Cultural Anthropology; it was also one of the issues addressed by a collective research project I was involved in, called *Diversities: A Project On People and Institutions*, sponsored by ENI's Enrico Mattei Foundation (FEEM), and carried out under the scientific supervision of Giulio Sapelli, Professor of Economic History.

I would like to thank ENI's Enrico Mattei Foundation and Professor Giulio Sapelli for supporting my research in India; Professor Marco Aime for letting me work freely on my own research project; the Human Sciences Research Methodology PhD teaching collegium at the University of Genoa's Faculty of Education Sciences; Professor Pinuccia Caracchi and, particularly, Professor Alessandra Consolaro, for sharing their profound knowledge of India with me; Professor Giorgio Solinas for the thought-provoking discussion during our PhD evaluation session; and lastly, Professor Giuliano Boccali. My thanks also to Carlo Petrini and Federica Tomatis for letting me know that Raghunath Medge and Gangaram Talekar, President and Secretary of the Nutan Mumbai Tiffin Box Suppliers Charity Trust, would be attending the 2006 "Terra Madre" event in Turin, thus providing me with the opportunity to meet them. My special thanks to all the *dabbawalas* who were patient enough to talk freely with me during our interviews.

I would never have been able to carry out this research without the help of many different people: I would like to thank Francesca Caccamo for translating my words into Hindi during interviews with non-English-speaking local informants; Usman Sheikh for interview transcriptions; Rebecca and Kenneth David for hosting me during my first stay in Mumbai; Shailindra Kaul and Abjijeet Sandhu for their whimsical yet important friendship; Sandy; Kalpana; Chiara Longo and her husband Sebastien Bastard for their hospitality during my second stay in Mumbai; Clara; Meena Menon; and the Annapurna Association. I would also like to thank Professor Roberta Garruccio, who—after reading and supervising my final dissertation at Milan University several years ago—also offered to comment on my doctoral dissertation. I am particularly grateful to the late Armando Marchi, head of Barilla-Lab, whom I remember fondly for the stimulating food industry research he allowed me to conduct at Barilla. My heartfelt thanks go to Daniele Cologna for all the advice that has been forthcoming over the years, accompanying my intellectual evolution.

I would like to thank Alessandra Tosi and Open Book Publishers for including my book into their catalogue, and my editors Corin Thorsby and Bérénice Guyot-Réchard for their comments and intellectual advice. My heartfelt thanks to Ishan Mukherjee, wishing him all the best for his studies. Finally, I would like to thank my translator Angela Arnone, with whom a close relationship developed as we debated the most appropriate route for rendering my research and ideas for the English-speaking reader.

This work is dedicated to Kenneth David, killed tragically in a fatal plane crash while I was in Mumbai.

And to my loved ones.

Preface

This book is about the anthropology of the city or, more accurately, anthropology *in* the city, based on the extensive map of one of the many systems of circulation: food. Food that is carried, delivered and returned from the kitchen to the consumer. The Mumbai *dabbawalas* are food deliverymen that connect homes and workplaces—messenger boys, urban servants who are fast and precise, trustworthy and discreet, clean and punctual. Service, certainly, but service immersed in the teeming ocean of urban modernity. Each day they move along the rail network; their work thus entails a journey and this journey is repeated on a daily basis, with long itineraries cadenced by the sequence of customer addresses where they must deliver without fail the tightly sealed tin that each wife has prepared and handed to them early in the morning, to be taken to a husband working in an office, on a construction site, in a shop, many kilometres away.

A mild sense of duty, of a delicate, humble and scrupulous mission, interwoven with a generous readiness to work for the good of the customer: these are the recurring motifs of work that seem to make the *dabbawalas* happy. They bring together the beneficiary and the benefactor (and is this not pure *Jajmani* philosophy?) in a shared satisfaction, yet seem to expand unexpectedly in the heart of frenzied modernisation. Food is a message, transmitted through nutrition: more than in other contexts, its energetic communication is released socially and physically in space. Born out of tender, loving care, it bridges the distance between one individual and another, passing the expanses desecrated by traffic, the mingling of people and vehicles, environmental impurities of exhaust gases, and inclement weather.

The custom of ordering takeaway food, to be delivered from the restaurant to the consumer's house, is far more widespread in the western world, although it is also to be found in Indian metropolises. This is a formula that every now and again replaces home-cooked food prepared in the family kitchen, like "going out to eat" without actually going out, a small exception to domestic routine. The *dabbawala* service is just the opposite or the reverse: it conjures up the feeling of home for those away from home. Each day it reinforces ties between the family and the workplace so that the domestic intimacy enclosed in the tiffin can emerge during a lunch break in the office, on a building site, in a factory. In this respect, it is quite similar to the custom once frequent in rural society whereby all kinds of farmworkers were ensured a midday meal.

Sara Roncaglia's description of the Mumbaite system reveals that, in contrast with more sophisticated market cultures, the order of affections and food containers maintains its tenacious hierarchy of precedence, which is as much about ethics as it is about taste and aesthetics. This is an order that establishes the indissolubility of the nutritional bond between family and work, men and women, etiquette and bodily ritual, and community membership. I believe that this is where the source of an investigative critique can be perceived, suggesting opportunities for research that will sound the innermost depths of the emergence or development of strongly cultural new urban trades. Such trades can take root deep in the cultural sensitivity of a society swarming with ethnic and religious contacts, innervated with open technologies and abysmal poverty, imbued with deep malaise and rocked by the tremors of social distinction. The Mumbai *dabbawalas* are not just a trade corporation but also a structured community, with dense social identity and cohesive recognition ties. The network of responsibilities, functions and organisational complementarity that forms the setting for the work of something like five thousand meal delivery men does not serve only to ensure the best technical standard for the service system. It could be likened to a modern guild, where work and social identity, devotion and economic gain, even the sharing of beliefs and religious works, as well as mutual aid, are part of this business culture.

Similarly, and from the same root, they produce and administer a symbolic substance without which the very existence of services would be compromised, or at least altered, in that implication of oriental *charitas* (and quite different from Christian charity) in which the welfare of the customer and the service provider are identified. This concept leads to a reflection on the comparison between the economic principles that distinguish different cultures: Italian, of British origins (slowly assimilated in Latin regions), and Hindu, or more broadly southern Asian. In the first case, it is the meeting of interests (or egotisms) to drive the motor of exchange and ultimately to fuel market solidarity on the basis of a useful cross-calculation: a concordance of convenience. In the second case, which extols trade and links the good to the useful, opportunities to achieve personal benefit is seen (or is represented) as an offering for the benefit of others: the offering of oneself or simply an offer to accord with the customer's contentment. In the most intense versions of devotion, indicated in traces of tradition, of *dharma*, this projective orientation achieves forms with greater signs of voluntary dependency. There is no need to stress that in this economic ethic the rhetoric of selflessness (of an uplifting mission) assumes the role of an ideology of social status and easily becomes an image—something that approaches advertising, the self-satisfied glorification of the corporate self, generous, benevolent, humble, and even joyfully submissive.

A fine, tenuous but persistent web enfolds the ramifications of a city that stretches endlessly, enfolds it with an artful ballet of deliveries, cadenced in minutely signed identification symbols (the *dabbawala* alphabet, writes the author: a system of distinctive symbols for groups and individual carriers, also designating places for sorting, delivery and destination). This sort of encrypted language, similar to an elementary information system that combines space and people, actually accompanies the daily weaving of the impalpable web of clientele and servers. Filaments of paths, competition, commodification: no less than other utilitarian-type exchanges, here the portions of comfort (perhaps consolation or affection) that the tiffin contains in its sealed interior, incorporate the insuppressible quality of the contents.

In the final part of her book, Roncaglia gives an overall (and wide-ranging) key for interpretation: gifts and merchandise move hand in hand in this system. Perhaps they even fuse, complement each other. In the wake of Godbout, the scenario transcends the cold mechanics of efficiency, profitability and monetisation: the ultimate utility of the cycle of patrons, services and remunerations does not drain away in the production or reproduction of material advantages, but in the creation of community ties. Compared to this encompassing aim, *dabbawala* work appears as a "business activity incorporated in a moral perspective". In general, this opinion can be accepted, provided that the commitment does not preclude further steps, which may be even more unpredictable and riskier, and may lead to the products of this moral economy flowing effectively into other, uncontrollable market circuits.

Pier Giorgio Solinas
Siena, 3 March 2012

Introduction

This book is an ethnographic analysis of a local workers cooperative in Mumbai: the Nutan Mumbai Tiffin Box Suppliers Charity Trust (NMTBSCT). This enterprise employs up to 5,000 *dabbawalas*, who have been delivering 200,000 lunch-*dabbas* daily to students, office workers and factory workers since the end of the nineteenth century.[1] A *dabba*, also known as a "tiffin", is a specially designed circular steel box made up of three separate sections that fit together to form a cylinder of about 20 cm in height. These food containers are commonly used by Mumbaikars (the inhabitants of Mumbai) to carry their lunch, which is prepared in their home and then delivered to them in their place of work by a *dabbawala*. The system allows everyone to eat home-cooked food without hygiene and cross-caste contamination risks.

The first chapter looks at the cultural, historical, and economic relationships between the city of Bombay-Mumbai and the NMTBSCT.[2] The city provides the dynamic backdrop for the establishment of a system of food distribution that offers a sustainable method of feeding the city in

---

1 I decided not to use diacritic marks when spelling Hindi and Marathi terms (nouns, names of people and places); nor do I use any Anglicisms in the transcription – such as double vowels (e, o) to express long vowels (i, u). The only exception is the term "*dabbawala*", formed by the noun "*dabba*" and the suffix "*wala*", which turns the word into a compound noun (like, for instance, "milk" and "milkman"). Please see the glossary for original spellings. The names of Mumbai districts are the official versions applied by the city's authorities.

2 Throughout the text I have attempted to follow a historiographical approach, referring to the city as Bombay when referring to its history up to 1995, when the name was changed to Mumbai, and as Mumbai when discussing its situation during the subsequent years. The name-change came about as part of a concerted government strategy to set modern India apart from its colonial past. Yet given how the city's inhabitants themselves tend to associate different meanings and allures to the old and the new name, in some cases I have found it more meaningful to keep the two names as one single construct (Bombay-Mumbai), reflecting two different, and yet complementary ways of understanding the city's complex soul.

harmony with traditional values. The *dabbawalas* do not consider this to be merely a job, a viable means for mostly poor and illiterate workers to survive: they see it as their profession.

The second chapter describes how religion, caste, and ideology have converged to generate meaning, ascribing specific values to Indian food. Here I apply a gastrosemantics-oriented approach, exploring how culture makes use of food to signify, comprehend, classify, philosophise, and communicate. This chapter offers a description of the complex relationships that link this process of cultural semantification of food to daily religious practices, the daily routine of Indian women and, lastly, surviving caste-related hierarchies in a vast Indian metropolis like Mumbai.

The third chapter describes the organisational structure of the NMTBSCT—its operational guidelines, its generational turnover, distribution logistics, the delivery process, and the technical solutions that make it extraordinarily efficient despite considerable odds. This includes simple techniques—like the symbols drawn on the *dabba* to identify the recipient's location—or more complex expedients, like the use of the railway network as a sort of mind-map that allows the *dabbawalas* to establish a symbolic and material affinity with this megacity of nineteen million inhabitants.

The closing chapter penetrates the tight-knit relationship that links the entire system of *dabba* preparation and distribution to the cultural processes of Bombay-Mumbai's nutritional transformation. The chapter traces this relationship back to the reasons that have made this Indian metropolis a truly global city; it looks at the eating habits and value systems ascribed to food by the many different migrant groups that make up the city's population. The ongoing acculturation process that accompanies the continuous inflow of migrants of very diverse origins has forged the city's characteristic nutritional physiognomy, recognisable in the diversity of cuisines and eating habits. Yet as the shift from old Bombay to new Mumbai progressed over time, there have also been changes in the tensions between different minorities and local communities, exacerbated by the city's growing ethnicisation. Certain groups have claimed collective rights on the grounds of identity and affiliation to particular castes, regional origins or language. Mumbai has become the stage for bloody racial and religious clashes, and the groups involved usually consider food the prime marker of differentiation and separation. Food has come to express distinctions and rivalries that to some extent already existed within the Indian cultural tradition, but have now been allowed to degenerate into overt political hostility and outright violence. In this harsh new climate, the "other" is subject to a kind of cultural cannibalism, as each social group aspires to an exclusive monopoly of power and culture.

These conflicts and changes are examined using the "foodscape" concept—a comprehensive approach to global symbolic and material shifts that affect food itself, food cultures and nutritional practices. The case of the *dabbawalas* helps us to understand how taste—the discerning and distinctive aspect of any food-related practice—is becoming a key factor in worldwide cultural transformation. Taste is not conceived simply as a sensorial impulse, but as a signifier, a cultural construct that is socially engineered to transform and lend new meaning to geo-political relationships.

Finally the appendix provides an extensive introduction to the fundamental issues that made my fieldwork possible. It analyses the polysemic nature of cultural diversity, embracing the multitude of meanings attributed to the subject. The diversity theme is usually addressed in relation to practices of social acceptance or rejection of otherness within organisations and institutions. In this perspective, my research is closely entwined with notions of identity, gender, and economic and social status in ethnic and religious minorities.

The book's title, *Feeding the City*, grew out of this consideration and the verb "to feed" is used here in the sense of "providing nutrition". It is an explicit reference to the way a nutritional regimen, a specific diet, affects an organism's state of good or poor health. Stretching the organic metaphor, food can be seen as a vector of phenomena expressing the easy or uneasy coexistence of different cultures in urban contexts. In this perspective, the way the city feeds itself is crucial for a broad cultural anamnesis of Mumbai. Thanks to the daily work of the *dabbawalas*, these cultural shifts come to light as the meals are ferried around the entire city in a distribution system that offers a tangible testimony of cultural coexistence mediated by one of its most potent signifiers, and the one most essential to human physiology: food.

As the twenty-first century ushers in an era of increasing anxiety with regard to humanity's ability to feed itself, we also witness the gradual global ascendance of a unified cosmology of tastes well as a heightened concern with nutritional practices. This trend is driven by a growing consensus on the importance of food—what it means, how it is produced and processed—and the deeper ethics of its preparation and consumption.

1. Bombay-Mumbai and the *Dabbawalas*: Origin and Development of a Parallel Economy

But if we do look back we must also do so in the knowledge—which gives rise to profound uncertainties—that our physical alienation from India almost inevitably means that we will not be capable of reclaiming precisely the thing that was lost; that we will, in short, create fictions, not actual cities or villages, invisible ones, imaginary homelands, Indias of the mind.

— Salman Rushdie[1]

Midday in Mumbai: teeming traffic besieges the city, lines of cars creep forward at a snail's pace, people walk in the road, buses swerve into their bays for a split second, rickshaws and taxis veer into every tiny space, while placid cows browse amongst all kinds of garbage. Hooting horns and chaos. Lunchtime is coming up for most civil servants, office workers, and school children. Nearly two hundred thousand people are waiting for their *dabbawalas*, who arrive promptly with the tiffins they have to deliver.[2]

---

1 Salman Rushdie, *Imaginary Homelands: Essays and Criticism 1981–1991* (New York: Viking, 1991), p. 10.
2 The term "tiffin" refers to a light meal popular during the British Raj. The word first made its appearance in the early 1800s and derives from the English verb "to tiff", referring to the consumption of a midday meal, and "tiffing", a slang term meaning the consumption of food and drink between meals. It survives in Mumbai's daily vocabulary to indicate a meal eaten away from home, as well as being used by the *dabbawalas* as a synonym for *dabba*. For further information, consult K. T. Achaya, *Indian Food: A Historical Companion* (New Delhi: Oxford University Press, 1994).

DOI: 10.11647/OBP.0031.01

Dabbas make a long trip every day to reach the people expecting them: a journey through the winding streets of this metropolis, with its twenty million or so inhabitants, and a solid history that goes back almost one hundred and thirty years.[3]

Origins of an alliance

The history of the *dabbawalas* runs parallel to that of Bombay itself. The archipelago that developed into the modern metropolis of Bombay became a centre of international trade during British rule.[4] The city was given to Charles II by the Portuguese as part of the dowry for his marriage to Catherine of Braganza in 1661. In 1668, the city was leased by the Crown to the English East India Company (operating at that time out of the port of Surat in present-day Gujarat) for ten gold sovereigns. It was not until about 1780 that Bombay began to exceed the importance of Surat, India's leading trading port. Thanks to exports of raw cotton and opium to China, what had appeared as a dreary fishing town—where the British had not expected to survive for more than two monsoons—became the second most important city of the colonial Empire.

The 1861 American Civil War gave further stimulus to Bombay's development as the British textile industry moved its bases to India and used the city as a production and export centre. The metropolis experienced startling economic growth and attracted significant amounts of capital for the creation of new investment and employment opportunities. The most evident aspect of this change, a trait of Bombay still seen today, was a migrant workforce arriving from outlying rural areas in search of employment. The gradual extension of roads and railways (the first railway line from Bombay to Thana was opened in 1853) made it easier for increasing numbers of people to travel all over India. The end of the

---

3   *Dabba* means "box" in Hindi. In this case it means a special container, made of steel and consisting of three separate sections that assemble into a cylinder about eight inches high, used specifically for taking lunch to work. The noun *dabbawala*, formed by the noun *dabba* and the suffix *wala*—which turns the word into a compound noun—means "he who carries *dabba*".

4   The city was named following the 1534 landing of Portuguese conquerors in the archipelago of seven islands known as Heptanesia (Greek for a "cluster of seven islands") in the Bronze Age. The islands of Bombay, Colaba, Mazagaon, Little Colaba, Mahim, Parel and Worli were called "Bom Bahia", the "welcoming port", by the Portuguese. When the city became part of British Crown possessions its name changed from "Bombaim" (the crasis of Bom Bahia) to Bombay.

American Civil War and the ensuing crash of cotton prices were the first stumbling block in the city's industrial expansion. But when the Suez Canal opened in 1869, it reduced the distance to London by approximately three-quarters, and cotton exports became one of the major contributors to the colonial economy. Bombay, a point where the land meets the seas, was christened *urbs prima in Indis* by the British and grew into a commercial hub for the whole of India.

The transformation from a fishing village to an important industrial city was partly the product of Bombay's connection to the British Empire. It actually became common to think of the city as the main driver of westernisation for the Indian subcontinent, although it was equally true that the centripetal forces moulding its commercial and industrial development were not just underpinned by western modernising forces. The Indian commodity market was linked to broader production and trade relations with the hinterland and with foreign markets (for instance trade in sugar, indigo and opium),[5] and its cotton mills relied on increased production and domestic market penetration. By 1920 Bombay held two fifths of India's total foreign trade, seventy per cent of coastal trade, and the majority of exports to the Persian Gulf and ports of East Africa. The city slowly evolved into a business hub, simultaneously turning into a political, administrative and educational centre where the arrival of new money created opportunities. It therefore attracted increasing numbers of migrants from all over India and the old continent, leading to the development of new forms of cohabitation and social organisation.[6]

A city of migrants

Bombay's remarkable development was reflected in the evolution of its social and demographic profile. In 1661, the population was estimated at

---

5 Giorgio Borsa, *La nascita del mondo moderno in Asia orientale. La penetrazione europea e la crisi delle società tradizionali in India, Cina e Giappone* (Milan: Rizzoli, 1977). Kirti Narayan Chaudhuri wrote that "The colonial impact on Asia was not confined just to diverting the flow of trade in a longitudinal direction from the previous latitudinal flow; it reoriented Asian intellectual thought in a similar direction as well". See Kirti Narayan Chaudhuri, *Asia before Europe: Economy and Civilisation of the Indian Ocean from the Rise of Islam to 1750* (Cambridge: Cambridge University Press, 1991), p. 11.

6 Rajnarayan Chandavarkar, *The Origins of Industrial Capitalism in India: Business Strategies and the Working Classes in Bombay, 1900–1940* (Cambridge: Cambridge University Press, 1994); Gillian Tindall, *City of Gold: The Biography of Bombay* (New Delhi: Penguin, 1982).

about 10,000; by 1872 it had risen to 644,405; by 1941 it was at 1,489,883.[7] A series of events were decisive for this population growth which included opium trade with China; the outbreak of the American Civil War; the expansion of the textile industry and the end of World War I.[8] Of course, there were also times when this steady flow of people dropped, in particular at the time of the 1918 famine and influenza epidemic, but it never stopped completely. If a city's vitality can also be seen in its ability to attract, then Bombay has certainly never ceased to be the destination for the dreams of millions of people. This progressive demographic increase became a growth pattern characteristic of the city, a model that formed an urban cultural landscape with a policy of being open to migrants from different contexts, welcoming and integrating faiths, languages, and ethnic groups.

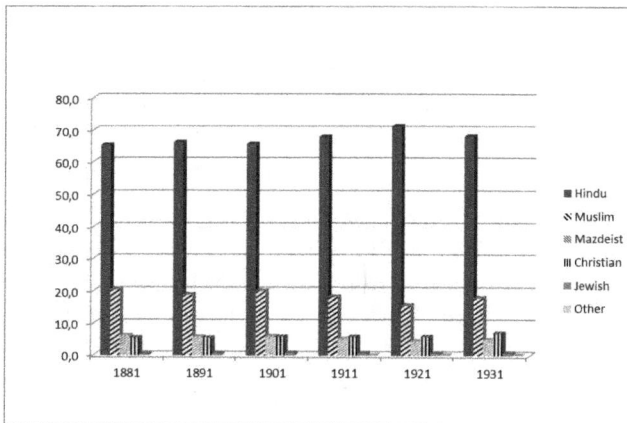

Figure 1. Percentage distribution of Bombay population classified by religion, 1881–1931.[9]

As is evident from the table above, in the years 1881 to 1931 the city was open to all types of worship but had a Hindu majority accounting for about two thirds of the resident population. Although not specified here, the

---

7   Chandavarkar (1994), p. 30.

8   Mainly thanks to the significant fortunes of the large Jamsetjee Jeejeebhoy Parsi and Sons mercantile agency, the leading exporter to China of opium produced in Malwa, Gujarat; see Peter Ward Fay, *The Opium War 1840–1842* (Chapel Hill: University of North Carolina Press, 1975).

9   Source: Rajnarayan Chandavarkar, *The Origins of Industrial Capitalism in India: Business Strategies and the Working Classes in Bombay, 1900–1940* (Cambridge: Cambridge University Press, 1994), p. 31.

category 'Hindu' embraces sister faiths like Jainism, Buddhism and Sikhism. Other religious groups were more or less minorities. The Parsees, for example, accounted for about five per cent of the population, but the role they played in trade and in business afforded them significant economic and political influence, despite their small number.

The Parsees originated in Persia and were descendants of the last Zoroastrians, migrating to India in the sixth century and settling in present-day Gujarat to escape religious persecution by the Muslims. Attipat Krishnaswami Ramanujan tells the story that, when the Parsees arrived in Gujurat, the region's ruler opposed their presence and sent them a diplomat holding a symbolic message: a glass filled to the brim with milk, indicating that the container could hold no more. The Parsees then sent the monarch back his full glass of milk in which a spoonful of sugar had been dissolved, expressing their intention to mingle with the native population as sugar does with milk: sweetly and taking up no space. The ruler was pleasantly surprised by this gesture and welcomed them.[10] In the mid-1600s the Parsees moved from Surat to Bombay because the British Governor, Gerald Aungier, offered favourable conditions to those who wanted to come to the city.[11] One of the main requirements for settling in Bombay was that they agreed not to preach their religion, a pact still respected by the descendants of the ancient Mazdeists.

The most important religious community after the Hindus was that of the Muslims, who made up about one fifth of the population. What the numbers do not reveal is that the Muslims (like the Parsees and Hindus) were and remain a heterogeneous group. Socially they are a stratified population of various sects: Shiites, Sunnis and Ismailis of the most diverse denominations. In Bombay there were two main groups: the Khoja and the Bhora.

---

10 Attipat Krishnaswami Ramanujan, "Food for Thought: Toward an Anthology of Food Images", in *The Eternal Food: Gastronomic Ideas and Experiences of Hindus and Buddhists*, ed. by Ravindra S. Khare (Albany: State University of New York Press, 1992), pp. 221–50 (p. 238).

11 The complex history of the Parsees very briefly outlined here can be explored fully by reading Eckehard Kulke, *The Parsees in India: a Minority as an Agent of Social Change* (New Delhi: Vikas, 1978), or the excellent two-volume work by Dosabhai Framji Karaka, *History of Parsees; Including their Manners, Customs, Religion and Present Position* (London: Macmillan, 1884). John Armstrong defines Parsee migration as an archetypal diaspora, because the Parsees have succeeded in safeguarding the bonds with their ancient myths and their distinctive alphabet. See John Armstrong, "Archetypal Diasporas", in *Ethnicity*, ed. by John Hutchinson and Anthony D. Smith (Oxford: Oxford University Press, 1996), pp. 120–26.

The Khoja were a caste of traders established in the fourteenth century by a follower of the Agha Khan, spiritual leader of the Ismailis sect. The term Khoja is the Indianised version of the Persian word *Khwajah*, meaning "respectable, rich person, wealthy merchant". Traditionally engaged in commercial activities, the Khojas are converted Hindus, who keep accounts in Hindi, and follow Hindu customs. In 1847, the Bombay High Court actually ordered that the Muslim law of succession was not to be applied to their communities. So, for instance, women are excluded from the right to inherit property. Moreover, the rules applied to marriage, divorce, birth and funeral rites are different, merging Muslim and Hindu practices.[12]

The Bhora, also known as "Bohara" or "Vohra", whose name derives from the Gujarati *vohorvu* or *vyavahar*, from the verb "to trade", are Shiite descendants of Hindus who had converted to Islam. The earliest communities can be traced back to Gujarat in the eleventh century and fall mainly into three distinct groups: Ismaili, Jafara, and Dawoodi. While the Ismailis swore loyalty to the Da'i Mutlaq in Yemen, the Jafara adopted Sunni Hanafi beliefs; after the schism, the Bhora Ismailis were heavily persecuted by local rulers. The Dawoodi, considered the best organised of the three groups, were the last to be formed by the two Da'i (the foremost being Tahir Sayf al-Din) and contributed to the shaping of the current community. The members of the Bombay Bhora community are chiefly small-scale itinerant vendors of bric-à-brac and trinkets or meat. Some became particularly wealthy by trading with China. As a consequence of this new-found wealth, some descendants of these families have had access to higher education, become judges or doctors, and are esteemed professionals in the city.[13]

Lastly, there are small communities of Christians and Jews who have distinguished themselves in the same way as the Parsees through the important role they have acquired in public and business life. Bombay's Jewish community is currently found mainly in Thane and it falls into three key groups: the Bene Israel (meaning "Children of Israel"), who are the most numerous and consider themselves the descendants of the first Jews who arrived in India about 2,000 years ago; the Malabar or Sephardic Jews, also still called "black Jews", whose ancestors came to India from Eastern Europe, Spain and Holland about 1,000 years ago, settling in Cochin; finally

---

12   Reginald E. Enthoven, "Kojah", in *The Tribes and Castes of Bombay*, ed. by Reginal E. Enthoven, 3 vols. (Bombay: Government Central Press, 1921), vol. 2, pp. 218–30.

13   Asaf A. A. Fyzee, "Bohoras", in *Encyclopaedia of Islam*, 12 vols. (Leiden: E. J. Brill, 1960–2005), vol. 1, pp. 1254–55.

there are the Iraqi Jews, called the "Baghdadi", who arrived in the late eighteenth century from Iraq, Syria and Iran, fleeing political and religious persecution; they expanded the trade network by setting up economic contacts with Singapore, Hong Kong, Kobe, Aleppo and Baghdad. The Bene Israel group is the biggest of the three Bombay groups and it built the first synagogue, Shaare Rahamim, in 1796. The community acquired particular prestige during the British Raj, when it emerged by developing its businesses and working for the British military corps. While maintaining eating (kosher food), religious (observation of the Sabbath) and hygiene (circumcision) practices typical of their faith, the Indian Jewish community has assimilated local customs and practices like language (predominantly Marathi and English) and the social caste divisions.[14]

The Christian-Catholic community has been present since the settlement of the first Portuguese in the seven-island archipelago, founded by the Franciscan friars who arrived on the ships coming from Europe. Historical evidence suggests that they landed as early as the first century AD, with St Thomas Apostle, who began his preaching from the southern coastal areas. The Syro-Malabaric church is one of Kerala's main Christian denominations and bases its liturgy on the Thomayude Margam (the law of Thomas). In the early period, Bombay Catholics soon built churches and monasteries, converting the local Koli tribes of fishermen.[15]

When the city was ceded to the British, missionary work was continued through the Church of Goa. The historical vagaries of this order are long and complex, and it is sufficient to remember that in 1720, members of the Goan clergy were expelled from the city for political reasons and the Vicar of Great Mughal (formerly Vicar of Deccan) was invited to protect the Catholic community with the Vatican's approval. Despite this, the Goan clergy always tried to recover its position within city government and in 1764 it established a "double jurisdiction" which took the name of the

---

14 See Nathan Katz, *Who Are the Jews of India?* (Berkeley: University of California Press, 2000).

15 The term *koli* actually means "spider" and in Marathi "the weaver of a web", a meaning derived from the work performed by this tribe. See Vinaja B. Punekar, *The Son Kolis of Bombay* (Bombay: Popular Book Depot, 1959), p. 5; Kavita Rane, *An Observational Study of Communication Skills Involving Fish Retailers in Mumbai* (unpublished MA thesis, University of Mumbai, 2005); and Sanjay Ranade, "The Kolis of Mumbai at Crossroads: Religion, Business and Urbanisation in Cosmopolitan Bombay Today", paper presented at the 17th Biennial Conference of the Asian Studies Association of Australia, Monash University, Melbourne, 1–3 July 2008, available at http://artsonline.monash.edu.au/mai/files/2012/07/sanjayranade.pdf [accessed 20 July 2012].

Vicariate of Bombay. Mumbai's Catholic community has numerous schools and non-profit charitable institutions that offer assistance to children, lone women and other vulnerable people. Its ethnic composition is quite varied because it includes Keralite, Goan, and Konkani Catholics (to mention just the areas of origin where the Catholic religion has its most massive presence), but also converted Kolis. Bandra is one of the areas with the largest concentration of Catholics in Mumbai, although they live all over the city.[16] Protestant history differs somewhat, because missionary activity was promoted by early settlers: to a lesser extent in the south by the Danish and Dutch; more intensely in the northeast and centre of India by British missionaries.[17] In Mumbai there are many Protestant churches, especially in the central part of the city, in the Malabar Hill district.

Like the Parsees, these small communities of Christians and Jews play the role of "middleman minorities".[18] This term refers to those ethnic groups—often immigrants or those arriving in the wake of diaspora dispersions—that occupy an intermediate position within the social structure, which allows them to play a role of economic intermediation between social entities separated by relatively strict status demarcations. It is no coincidence that in history the middleman minorities emerged mainly within strongly segmented feudal societies (like Jews in Medieval Europe or Armenians in the Ottoman Empire) or those based on castes (for instance Parsees in India).

---

16   Sebastian Irudaya Rajan, *Catholics in Bombay: A Historical-Demographic Study of the Roman Catholic Population in the Archdiocese of Bombay* (Shillong: Vendrame Institute, 1993); and Felix Alfred Plattner, *Christian India* (London: Thames and Hudson, 1957).

17   Antonio Armellini, *L'elefante ha messo le ali. L'India del XXI secolo* (Milan: Egea, 2008), p. 151.

18   The "middleman minority" theory was developed in the United States by scholars of immigrant socioeconomic integration strategies, in particular Edna Bonacich, "A Theory of Middlemen Minorities", *American Sociological Review*, 38 (1973), 583–94; and Edna Bonacich, "Middleman Minorities and Advanced Capitalism", *Ethnic Groups*, 2 (1980), 311–20; see also Walter Zenner, "Middleman Minority Theories: A Critical Review", in *Sourcebook on the New Immigration*, ed. by Roy S. Bryce-Laporte, Delores M. Mortimer and Stephen Robert Couch (New Brunswick, NJ: Transaction Books, 1980), pp. 413–25. There should be an element of caution with regard to the original middleman theory. This "classic" view of middleman minorities tends to attribute a concrete nature to "cultural" elements of circumstances and provisional condition, like the sojourner status that is often the outcome of explicit exclusion policies or non-recognition of resident status/citizenship. The intention was useful but insufficient and not appropriate for grasping the complexity of the dynamics that today characterise the realities of minority immigrant entrepreneurs in the societies they enter. See also Ivan Light and Edna Bonacich, *Immigrant Entrepreneurs: Koreans in Los Angeles, 1965–1982* (Berkeley: University of California Press, 1988).

These minorities also succeeded in societies characterised by stable regimes in which the dominant class was separated from the subjugated class (as in colonial societies like the Chinese in the East Indies), or those marked by enduring implicit statutes of subordination, as was the post-colonial case. These societies typically have strong socio-economic polarisation linked to labour market segmentation and to cultural barriers that discriminate against particular members of society for ethnicity, age, religion, etc.

Although middleman minorities are considered by the ruling classes as "foreigners who cannot be integrated" — and have often internalised their own "otherness" — they differ from other ethnic minority groups because of their role as providers of financial and business services. They enjoy a status midway between that of a ruling class and a subordinate class. Acting as mediators between producers and consumers, employers and workers, owners and renters, the elite and the masses, they bridge the status gap between the dominant and the dominated. However, the social buffer role played by middleman minorities is conditioned by the fact that they are still socially and symbolically vulnerable. This allows the elite to channel the resentment and hostility of subordinate classes in their direction during periods of heightened social conflict. In turn, this not only reinforces the ethno-religious and linguistic-cultural self-referencing (real or perceived) of these middleman minorities, but also their dependence on the stability of the predominant social structure.

Initially, this "tendency to self-reference" was traced back by Edna Bonacich to a cultural orientation and social planning typical of the sojourner (in other words the diaspora migrant, the temporary resident, perpetually seeking a future return to the homeland left or lost). Several scholars of migrations now attribute it to the complex dynamics of interaction that middleman minorities develop with both the ruling and subordinate classes, especially during or as a result of social conflict. The precariousness of their social status often leads these groups to prefer self-employment and work involving business and finance, which ensure strong capital liquidity and mobility. There is also a preference for building communities that to the outsider may appear to have a strong internal solidarity and capacity to resist assimilation. The cohesion of these communities may also be strengthened through endogamic marriage strategies, policies and practices for safeguarding distinctive cultural traits, as well as forms of residential, community and other sorts of segregation.

Mumbai's cultural and social ethnic stratification is reflected not only in the diversity of religions practiced in the city, but the multitude of languages spoken.

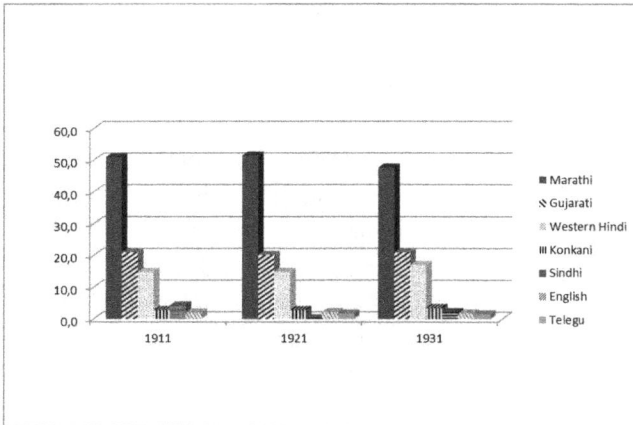

Figure 2. Percentage distribution of population classified by language spoken, Bombay, 1911–1931.[19]

Although half of the inhabitants speak Marathi, no ethnic group is linguistically dominant in social and business interaction. The constant arrival of new migrants increased linguistic complexity and led to the evolution of a *lingua franca* or creole known as Bombay or Bambaiya Hindi. Although this vernacular constantly changes, it is still spoken because of its use in 1950s and 1960s Bollywood films.[20]

Bollywood is the quintessential form of Indian film-making not only due to the large number of films made each year in Bombay, but because Bombay film companies release their films in Hindustani (a pluricentric language made up of Urdu and Hindi). Hindustani is not restricted to a particular area: it is spoken across India alongside other local dialects and was the vehicle that carried Bombay-made films across the nation. Although the arrival of sound forced production companies to make films in the different languages spoken on the Indian subcontinent, with Hindustani it was possible to reach a wide and varied audience.

---

19   Data: Rajnarayan Chandavarkar, *The Origins of Industrial Capitalism in India: Business Strategies and the Working Classes in Bombay, 1900–1940* (Cambridge: Cambridge University Press, 1994), p. 33.
20   For further information on Indian films, see Elena Aime, *Breve storia del cinema indiano* (Turin: Lindau, 2005), p. 83.

Bombay's human, cultural, and linguistic diversity made it a driver of initiatives for evolving a method of communication accessible to all its inhabitants.

Understanding that Bombay was—and still is—a city of immigrants is crucial to interpreting its cultural and social stratification. Most people who came to seek employment in the city maintain a close bond with their place of origin. Work in the city has typically been irregular and employment conditions often precarious, so ties with the rural homeland are common, serving as a safety net in the event of unemployment. The bonds maintained with these rural areas were a transitional stage in the formation of the urban labour force in the early period of Bombay's industrialisation, but were also a basic risk management strategy. Research undertaken in 1970 indicates that there was a steady integration process for migrant labour reaching the city, so this is a constant of Bombay life given that the active population is still composed mainly of migrants.[21] The city's unique urban format is underpinned by the fact that the work situation in Bombay is precarious and erratic, and that the labouring classes maintain ties with their place of origin. These villages are both a stable source of recruitment for cheap labour and a kind of outlet valve for excess labour in times of crisis. Many urban workers regularly send home money or goods, thus holding onto the status and rights acknowledged for their position in the family.

These migrants support themselves and their families by working as "coolies" in offices and homes.[22] At the time of the consolidation of the British Empire, the term "coolie" was used by the British to identify low-skilled employees, often bound by multi-year contracts for forced wage labour (indentured workers) in the colonies. The term was soon adopted

---

21 Kunniparampil Curien Zachariah, *Migrants in Greater Bombay* (London: Asia Publishing House, 1968).

22 The etymology of the English word "coolie" is uncertain. The most likely derivation is from the Hindi/Urdu word *qūlī* (a labourer) or the Tamil *kuli* (a wage), but some think that it comes from the Turkish word *kuli* (a slave). One last theory is that it originates from the name of an aboriginal Gujarat tribe, the Kuli or Kholi, subdued and forced into menial jobs by the British after the foundation of their commercial ports in Surat and Bombay. See Rana Partap Behal and Marcel Van Der Linden (eds.), *Coolies, Capital, and Colonialism: Studies in Indian Labour History* (Cambridge: Cambridge University Press, 2006); Daniele Cologna, *Cina a Milano* (Milan: Abitare Segesta, 2000); and Yann Moulier Boutang, *De l'Esclavage au salariat: Economie historique du salariat bridé* (Paris: Presses Universitaires de France, 1998). See also Marina Carter and Khal Torabully, *Coolitude: An Anthology of the Indian Labour Diaspora* (London: Anthem Press, 2002). The word "coolie" is still used to mean those workers who perform heavy duties, like porters who carry luggage at train stations.

by all European powers engaged in colonial expansion. In 1833, the British Empire abolished the slave trade, before abolishing slavery in its colonies in 1874; from then on, Indian and later Chinese coolies became the equivalent of African slaves for European industrial imperialism. They were essential for the development and commercial exploitation of the colonies. The term coolie indicated an unqualified, waged, unskilled worker and could also be used to some extent to describe a cotton mill worker, although this group usually retained links with farm work in their villages. So, despite having moved their permanent residence to Bombay, they acquired a different social status.

Clearly the term "model" applied to Bombay cannot be taken in its usual meaning of a "grid" required to explain a particular circumstance, as this would attribute an unrealistic uniformity to the Bombay migratory trend. Emphasis should fall on the huge differences from one region to another, and even amongst the individual villages of origin, the prevailing climate, nature and conditions of work. Whatever their origin or reason for leaving their home, almost every migrant in India preserved ties to their village and caste. The interaction between those rural bonds and the city's economic structure has fostered the development of a unique Bombay culture founded on practices, languages and expressions rooted in ethnic, class and caste origins, and a process of constant transformation.

From the legend

This ever-shifting social and economic landscape was the destination of Havji Madhu Bacche, a young man from the Marathi-speaking district of Pune.[23] Towards the end of the 1880s, an anonymous Parsee banker employed Bacche to go to his home in Grant Road, collect his tiffin and deliver it to his office on Ballard Pier. The young man was one of the many Maratha workers in *topi* caps who could—and still can—be seen

---

23 There are few relevant bibliographic sources and in most cases they refer to various interviews given by the president of NMTBSCT, Raghunath Medge. A number of researchers look at the figure of Bacche: C. S. Parekh, *The Dabbawallas of Mumbai* (unpublished PhD thesis, Narsee Monjee College of Commerce and Economics, Mumbai, 2005); Ramasastry Chandrasekhar, *Dabbawallahs of Mumbai*, Richard Ivey School of Business, University of Western Ontario, 2004, available at http://beedie.sfu.ca/files/PDF/mba-new-student-portal/2011/MBA/Dabbawallahs_of_Mumbai_(A).pdf [accessed 28 October 2012]; and a manuscript I was given by Medge, written by an anonymous author, with the title: *A Visionary Who Created History Through the Dabbawallas.*

at any crossroads, waiting to be hired for all sorts of work. This was the beginning of a legend:[24] the creation of a food distribution system that has progressively increased its catchment area. Raghunath Medge, president of the *dabbawala* association Nutan Mumbai Tiffin Box Suppliers Charity Trust (NMTBSCT) tells this story:

> It was 1890 when the system started, when the British were colonising India and Bombay was starting to be very well-off. There were railways, roads, post offices, government offices but people had to eat 'fast food', so Parsee women began cooking food as a business, like a canteen. They would cook any type of food people wanted, because in Bombay there are all sorts of food: Gujarati, Punjabi, Rajasthani, Maharashtra, vegetarian, non-vegetarian... really all kinds of food. Unskilled workers could be found at any crossroads, sitting there in their *topis*. One day a woman said to one of them 'Go deliver food from here to there' and 'I'll give you work, I'll pay you'. And he replied 'I'll do that'. At the beginning he carried about twenty, twenty-five tiffins, and he'd take them from Girgaon to VT Station.

Certainly Bacche's vision was crucial in establishing the regular group of workers, because he soon realised that his capacity to increase this delivery service relied on his ability to find people who could carry heavy wooden *dabba* trays. What seemed like typical "coolie" work soon took on a different connotation to straightforward delivery work, thanks to the formation of an association governed by a set of internal rules and with a solid reputation for reliability. Bacche's son describes his father's personality and professional ethics, and how he founded the *dabbawala* association:

> I remember that the *dabbawalas* got started like this: at that time my dad had gone to Bombay and was delivering food as a job. Other people already did this and he set up a union of all these people.[25] He collected money from all the people in the union and created *puja* [worship] for Satyanarayan [the "true Narayan" or "the true God", a name given to Vishnu/Krishna]. He then called a meeting, inviting all those who delivered food and he collected

---

24 Giulio Sapelli points out that if "the person in question has died, only *entelechiale* traces remain, which is to say the 'footprint' left by their work. This trace is the course of Jungian individuation, the gradual emerging of the 'silhouette' of an intention and a representation that is increasingly precise and clear". See Giulio Sapelli, "Mitobiografia per le scienze sociali", in *Giannino Bassetti: L'imprenditore raccontato*, ed. by Roberta Garruccio and Germano Maifreda (Soveria Mannelli: Rubbettino, 2004), p. 261.

25 Testimonies are unanimous in affirming that the association started up in 1890, but also that the *dabbawala* service already existed in Bombay in a non-organised form. The date must therefore be considered as the moment in which Bacche began this type of work, not as the moment the service was invented.

money from the *mukadam* [the group leaders], so they could buy a place to build the first *dharamshala* [a resting place for pilgrims] in Bhimashankar. He also invented a series of rules, for example that you couldn't take another *dabbawala*'s *dabba*. Then he occupied the baggage compartment on the train, saying that it was for our *dabbawalas*. The union continued this way. My father believed that everyone had to be fine. He listened to everyone's opinions, then acted for the good of all. My father did good.

Bacche acknowledged that one of the most important and most immediate strategic issues was to identify the resources for implementing the delivery service in the best way. In order to manage a group organically, he needed to know the people, speak the same language, and share the same relational code; hence recruiting his fellow villagers seemed the fastest and most logical method.[26] To this day, the *dabbawalas* continue to arrive from the same rural area in the Pune district, about three hours from Mumbai by train. The area includes small villages like Audar and towns like Rajgurunagar. It is a chiefly agricultural region and the uncertainty of work in the fields continues to drive people to go and seek paid employment in Mumbai. These migrants were (and are still) largely illiterate and therefore destined to expand the ranks of Mumbai's unqualified labour, so a chance to deliver tiffins was considered more desirable than agricultural work.

Bacche's chief merit was his creation of an organised working group, which then served as the foundation for the NMTBSCT. The sheer simplicity of the idea — a service for transporting food prepared at home by the family to a customer's place of work — underlies an entrepreneurial strategy based on the ability to exploit the interaction with Mumbai's complex ethnic and social configuration. The success of the NMTBSCT is based on developing trust between the *dabbawala* and the customer, on the ability of the work group to deliver lunch on time, on the excellent reputation for reliability and punctuality and, finally, on the organisation's ability to interact with the city's cultural fabric. Direct testimonies of customers of the past are not available, but a reconstruction of their profile indicates the precious climate of trust in which the *dabbawala* once worked, and still does.

---

26   The lack of direct testimonies makes it impossible to express considerations on the premeditated nature of this decision. The sociology of migrations and ethnic business studies suggests, however, that compatriot ties are often a very valuable ethnic resource for anyone setting up a business. See, for instance, Light and Bonacich (1988); and Alejandro Portes (ed.), *The Economic Sociology of Immigration: Essays on Networks, Ethnicity, and Entrepreneurs* (New York: Russell Sage Foundation, 1995).

Bacche's son, who is also an NMTBSCT *dabbawala* and a retired NMTBSCT director, describes how the meal delivery system began:

> I can't remember the name of the Parsee woman who gave my father the job, it was a long time ago. But this Parsee lady called Bacche and he started working. He wasn't the only one. It was a group: one person alone couldn't have delivered tiffins all over Bombay. The group started with fifty-sixty people. My father started in 1930 and Bacche in 1890. There were other people with him. Bacche died in 1955 and my father in 1980. Bacche's brother's son lives in the village and my house and my land are there; so is my mother. The *dabbawalas'* first customers were Parsees, Christians and Gujaratis who worked in the offices, in the Girmi cotton mills. Our salary in 1940 was twenty rupees per month, yes, twenty rupees. Our wages increased gradually. At that time the British were in charge and in Bombay there were eighty-four cotton mills in different areas: Girgaon, Dadar, Mahalaxmi, Grant Road. We had a lot of customers in the cotton mills and that was where I used to deliver the food. When they closed down work didn't fall off because the cotton mills became business centres. But that happened after many years, after ten years. We would collect tiffins from restaurants, Parsee *dhabas*, a very cheap sort of restaurant, and we'd deliver them to customers. Some were Christian, others vegetarian and non-vegetarian, some were Gujarati.

Figure 3. Picture of Madhu Havji Bacche. By kind permission of Raghunath Medge.

Serving the Bombay elite

The first customers for the *dabbawalas* arrived in the late nineteenth century and were largely members of the British and Indian elite. There were few European women in the British community able to cook western meals, and homes often had several chefs who cooked western and Asian-style food.[27] The British adapted to food in India in increments: there was a first phase during which Indian food was appreciated by British travellers, but was slowly replaced by a mixed cuisine with its own recipes, called Anglo-Indian food.[28] This cuisine is also the result of several cross-cultural marriages between Europeans and Indians. The term "Anglo-Indian" initially referred to British residents in India but was later used to denote the offspring of mixed marriages where there was usually a European father and an Indian mother. The difficulty in adapting to local eating habits encouraged Europeans to use the services of some Parsee kitchens on those occasions when they could not eat at home, and they would request for meals to be cooked in a way resembling the recipes of their homeland.[29] In addition, Bombay's poor hygiene conditions and chronic lack of drinking water made it even more difficult to adapt to the city's cuisine.

The Parsee community had an important role as culinary middlemen between the British colonial class and the city's immigrant populations.[30] As is typical with middleman minorities, especially those who have been persecuted in the past, many Parsees became bankers, merchants or doctors. Industrial and social change in Bombay provides ample proof that Parsee and British cultures harmonised, not least of all because the Parsees adopted English as their preferred means of communication.[31] To understand the importance that the Parsees acquired in the business field, it is useful to remember that they were the only non-Europeans to be shareholders in the Bank of Bombay, the Bank of Bengal and the Bank of Western India.[32]

The less evident side of this preferential relationship between European and Indian society (although it is the most important for this study) was the

---

27   See Tindall (1982).
28   For more information on Anglo-Indian cuisine and accounts by British travellers on their relationship with Asian food, see Achaya (1994); and Patricia Brown, *Anglo-Indian Food and Customs* (New Delhi: Penguin, 1998).
29   This information comes from various conversations with Raghunath Medge.
30   See Kulke (1978).
31   Ibid.
32   Borsa (1977), p. 139. For a more recent historical reconstruction, see Michelguglielmo Torri, *Storia dell'India* (Bari: Laterza, 2000).

creation of kitchens, generally managed by Parsee women. These women used their husbands' networks of acquaintances to provide lunches for Bombay's upper middle classes and elite. What made Zoroastrian cuisine a major medium of cultural exchange was its extreme adaptability to Hindu influences. Parsee cuisine actually has few dietary restrictions and reflects both its Persian heritage with its strong non-vegetarian component, and the adaptation of this cuisine to the dietary habits acquired during the initial settlement in Gujarat, a region that is prevalently vegetarian.[33] This positive fusion was particularly suited to the taste of the Europeans and Indians living in Bombay.[34] In the early twentieth century, the city was the common denominator in the different lifestyles that periodically came together.[35] It progressively revealed itself to be a place where every language in the world was spoken and where everything was eaten, with a healthy appetite. Medge describes Bombay's culinary cultural fusion:

> Mahajiraoji Bacche was the founding member of the *dabbawalas*, starting a small family business in 1890. Then there was my father and seven or eight other people. But the *dabbawalas* didn't cook the food. The Parsees began cooking in canteens. We started this system, we *dabbawalas*. When the British Raj began to develop Bombay, it built railways, post offices, stations. At that time there were no restaurants in Bombay, which was a commercial city and still is. Then came the French, the Dutch, the Portuguese, the Russians. People from all over the world settled in the city to work. Since masala spice mixture has cloves, cinnamon and many other ingredients that are not grown in India, those people came here and brought those spices to sell. At that time there were British government rules, people worked in offices, but there were no restaurants. Then we started to sell food from canteens. When I say we, I mean our family. We are an extended family, all farmers, living in hilly areas. We weren't rich. We couldn't read or write. Back home

---

33 The teachings of Mahavira, the master who founded the Jain religion in the sixth century BC, and of other Jain scholars of the eleventh and twelfth centuries generated a huge following for vegetarianism in Gujurat. Their doctrine was reinforced by the disciples of Vishnu with similar principles. About 70% of the Gujarat population today is thought to be vegetarian. See K. T. Achaya, "In India: civiltà pre-ariana e ariana" in *Storia e geografia dell'alimentazione*, ed. by Massimo Montanari and Françoise Sabban, 2 vols. (Turin: Utet, 2006), pp. 144-52.

34 Many other types of cuisine developed following European colonisation of India and the coming together of all these cultures. For instance, Goan cuisine from Portuguese colonisation; Pondicherry cuisine from French influence; and Keralite cuisine from a widespread Catholic influence.

35 The reference to the "common denominator" is from Mark-Anthony Falzon, *Cosmopolitan Connections: The Sindhi Diaspora, 1860–2000* (New Delhi: Oxford University Press, 2005), p. 101. The author defines Bombay as playing a connective role in the melting pot concept, which arose and evolved mainly in the United States.

the extended family just took care of the land: farming, livestock, some sold milk and dairy products. Everything depended on the rainy season. The rest of the people, who didn't have a job, came to Bombay to earn money, because if it doesn't rain, there are no crops.

Unfortunately, there is not a great deal of literature on the evolution of Parsee cuisine precisely because this invisible daily art was performed by women, and the recipes were handed down orally, from mother to daughter.[36] Nevertheless, it would be wrong to think that in the organisation of Bombay's food practices the role of women did not act as the focal point for a wider overview of food.[37]

... and the cotton mill-workers

The profound urban changes in Bombay also led to an increasing demand for the *dabbawala* service. Merchants selling products like diamonds, gold and clothing became an important part of the association's clientele. However, it was Bombay's skilled cotton workers who used the meal delivery system the most. From the construction of the first mills in the mid-1800s to their closure in 1980, Bombay-Mumbai bore witness to the changes brought about by the growth, expansion and decline of cotton manufacturing. The mill-workers made up a complex ethnic group, comprising mainly men arriving in the city from outlying villages in search of work. Through family, caste and shared geographical origin, they found employment as factory workers and supported the family back in the village.[38]

---

36   For an explanation of many Parsee recipes, see Niloufer Ichaporia King, *My Bombay Kitchen: Traditional and Modern Parsi Home Cooking* (Berkeley: University of California Press, 2007). Parsee food traditions are also explained in Bhicoo J. Manekshaw, *Parsi Food and Customs* (New Delhi: Penguin, 1996).

37   Historian Caroline Walker Bynum has already highlighted the close relationship between women and food, especially in European medieval religious history. Widening the field of inquiry it can be seen that men may be involved in the production of food but it is usually women who convert the food into meals and women who control the household. The scholar's considerations can be applied appropriately to other contexts. See Caroline Walker Bynum, *Holy Feast and Holy Fast: The Religious Significance of Food to Medieval Women* (Berkeley: University of California Press, 1987). For a study of the relationship between women and food in more recent times, see Maria Guiseppina Muzzarelli and Fiorenza Tarozzi, *Donne e cibo. Una relazione nella storia* (Milan: Bruno Mondadori, 2003).

38   See Chandavarkar (1994); Meena Menon and Neera Adarkar, *One Hundred Years One Hundred Voices: The Millworkers of Girangaon: An Oral History* (New Delhi: Seagull Books, 2004); Jan Breman, *Of Peasants, Migrants and Paupers: Rural Labour Circulation*

More often than not, these workers' private lives were spent in small, crowded rooms in a building where they slept, called a *chawl*; meals were taken in small eating places called *khanawals*. These places were run by widows or women who had to support their families when their husbands were unemployed, and offered a dignified alternative to prostitution. Bombay's female workforce was, for the most part, not involved in factory work. The city was primarily home to male immigrants: in 1864 there were about 539 women for every thousand men, and the ratio was virtually the same in 1921 (about 525 per thousand).[39] This gap between women and men was closely connected to the migration model prevalent in Bombay: the precariousness of working conditions forced men to leave their families back in the village, returning periodically to work in the fields or when they lost their city jobs. It was only with the expansion of the cotton industry after 1880 that the demand for female labour grew substantially, but subsequent enforcement of maternity laws and limited working hours for women and children meant that female recruitment once again decreased.[40] This gradual detachment from the industrial world relegated women to occupations in domestic services, catering, small-scale sales or prostitution. It was mainly widows who fell prey to the latter, as they were victims of a strong social stigma and often had no access to other sources of livelihood.[41] Female unemployment was also the expression of asymmetrical gender relations in India. For this reason, these small eateries were very important for women, giving them the opportunity to secure a livelihood and avoid destitution.[42]

---

and Capitalist Production in West Bengal (Oxford: Oxford University Press, 1985); and Jan Breman, *Labour Migration and Rural Transformation in Colonial Asia* (Amsterdam: Free University Press, 1990).

39   Chandavarkar (1994), p. 94.

40   J. C. Kydd, "The First Indian Factories Act (Act XV of 1881)", *The Calcutta Review*, 293 (1918), 279–92; and Alexander Robert Burnett-Hurst, *Labour and Housing in Bombay: A study in the Economic Condition of the Wage-earning Classes of Bombay* (London: King & Son, 1925).

41   Deepa Mehta's film *Water* (2005) describes the condition of widows in traditional Indian society. Much has been written on the status of widows in India and it is worth remembering that until a few years ago, and this is still so in some parts of India, widowed women were blamed for not safeguarding the life of their husband, either through neglect or through bad *karma*. A widow with no adult sons was the living example of the precarious inequality between men and women. The position changed with the presence of a son, which allowed the woman to be considered the honoured mistress of the house. See David Smith, *Hinduism and Modernity* (Oxford: Blackwell, 2003).

42   For a history of women in the cotton mills, see Menon and Adarkar (2004). For more on *khanawals*, I recommend Dina Abbott, "Women's Home-Based Income-Generation as a Strategy towards Poverty Survival: Dynamics of the 'Khanawalli' (Mealmaking) Activity of Bombay" (unpublished PhD thesis, The Open University, 1994).

The administrator of the Annapurna Mahila Mandal, an association for battered women, describes exactly how the catering service has been a source of redemption:

> The Annapurna Mahila Mandal women's association was founded on 8 March 1975 by Prema Purav, on the occasion of International Women's Year. Prema Purav's life was fully committed to social action. She was born in Goa, in the small village of Khodeye, and she soon began working and helping her brothers, patriotic fighters against the Portuguese, taking food to the forest for them. She arrived in Bombay in 1953 where she completed her education and worked as a social activist for women's rights. Her involvement in the millworkers' union was crucial for understanding the often difficult conditions of women working in factories. In fact, most of the women who came to Bombay had no education, were in low-skilled jobs and were not protected by any maternity laws, so when they had children to raise they lost their jobs. The owners of the cotton mills would send them away and they became unemployed. In 1975, about seventy-five thousand women were sent home from their cotton mill jobs. The situation was further aggravated with the great strike of 1982, during which many workers lost their jobs. They had no education, so what could they do? Many had come from a village where they had left families and had to send money home. Women who had followed their husbands to Bombay and were employed as factory workers began to earn money to support their families doing what they knew best: cooking. That was how the *khanawals*—small, family-run kitchens—developed, offering simple, cheap food for workers who couldn't afford to eat at a hotel or a restaurant. Women cooked, selling and sending the food to workers coming from all the Maharahstra districts, and to street vendors. Certainly not all unemployed women began to do this sort of work, some returned to the village, others decided to organise themselves, despite being illiterate. But to start any sort of business they needed initial capital. Unfortunately they had no guarantees to offer a bank and so they turned to private moneylenders. Interest was very high and the women began to be exploited sexually, psychologically and financially by these loan sharks. In 1975, Prime Minister Indira Gandhi began to change bank loan laws and this allowed women to start businesses. Loans from banks had interest rates of four per cent per year: moneylenders wanted twenty per cent a month. It was precisely Prema Purav, a woman who had contacts with unions, who went door to door to inform people of this opportunity. So they began borrowing money from banks and formed a group, an organisation. As time passed, the organisation was given the name 'Annapurna', which means 'Goddess of Nourishment', because it was thanks to cooking that many women were able to escape difficult social and family situations.[43] And our customers are no longer just factory workers. We now serve offices, banks and schools.

---

43   In Hindu mythology Annapurna is the goddess of food and nourishment. She is also considered the goddess of prosperity and abundance. In Sanskrit *anna* means "grains"

The association is still present in Mumbai and has expanded its range of action, building refuges for jobless women in different parts of the city. The administrator of Annapurna Mahila Mandal, explains that it is also connected to *dabbawala* work because customers often use *dabbawalas* to transport food cooked by Annapurna:

> Mumbai office workers who cannot take food with them to work employ *dabbawalas*. As long as the *khanawals* existed, women cooked and carried the food, but after the foundation of Annapurna, customers began to send *dabbawalas* to pick it up. The association is not able to deliver very far away in Bombay, so the *dabbawalas* do it instead as they have a broader distribution network. Annapurna does not have its own *dabbawalas*: people who want its food send a *dabbawala* to pick it up and take it to their office. A similar structure now exists in Dadar (New Bombay) and the employees of several companies like Philips, Godrej and Tata order food, cooked without caste or religious distinctions. A number of factories have now been moved to the coast and most textile factories have shut down. Many workers who lost their jobs started small businesses, like greengrocers, and to help out with their financial needs the association set up a sort of micro-credit system to allow them to stock and start their businesses, thus avoiding the risk of borrowing from loan sharks. Some use the credit to educate their daughters. A third site is Navi Mumbai, where there is a multipurpose women's rehabilitation and health centre, with hostel facilities, now with more than fifty residents. Orphanages care for children up to the age of seventeen but when they have to leave they are without protection or work, so Annapurna adopts and trains them to work, as well as arranging weddings.

In the past, the food brought from the *khanawals* was based on very simple regional recipes and sold at extremely low prices. "Eating out" was not a special event for workers but a choice dictated by daily survival: houses were small and crowded and cooking was almost impossible.[44] Only the

---

and "food" but also "body/physical/shell", and *purna* means "full", "complete", "perfect". Annapurna gives food and she is also the goddess of the harvest, the protector of the fields so she is worshipped in order to have a good harvest. She is the goddess of the kitchen and she gives to the poor. She is the generous one (*puma*). Traditionally she is depicted with a container of food in one hand and in the other a spoon that she holds out to the faithful. She also feeds her husband, Shiva, putting food in the human skull he uses as a bowl. As goddess of food she also transforms food into energy. The food she gives to Shiva actually gives him the energy (*shakti*) for achieving his wisdom and enlightenment. In the same way that Annapurna symbolises the divine aspect of nourishment, through food the cook not only feeds the body but also provides the energy for those who eat to follow their own destiny. If the food is prepared following a sacred ritual, an outright alchemy is created. For this reason, images of Annapurna are found in kitchens, next to canteen tables and in restaurants.

44   Frank F. Conlon, "Dining Out in Bombay/Mumbai: An Exploration of an Indian City's Public Culture", in *Urban Studies*, ed. by Sujata Patel and Kushal Deb (New Delhi: Oxford University Press, 2006), pp. 390–413.

middle class could afford to eat at home and share the meal with their families, while the working classes used the *khanawals*. These places became real centres for socialisation, where it was possible to get together and talk, and where people could order tiffins to take to work. *Dabbawalas* followed the schedule of the cotton mills, delivering tiffins for each of the three factory shifts: from eight in the morning to two in the afternoon; from two to eight in the evening; and at night, from eight to one in the morning. There were about twenty *dabbawalas* working in Girgaon, Mumbai's industrial district, and each *dabbawala* managed forty customers.[45]

Interviews reveal the intense relationships that the *dabbawalas* built up with the millworkers. Medge observes:

> The Girgaon cotton mills were big factories. There are still some in Girgaon and Dadar. The Kohnoor, Down and Poddar cotton mills in Dadar are now shopping centres and offices. At the time of the cotton mills, we had plenty of customers. We delivered tiffin to three shifts. Morning, afternoon, and night shift. During the strike the cotton mill was shut down. Workers lost their homes because many of them lived in the mill grounds, so they went back to the village. When the cotton mill went on strike, workers didn't have the money to keep their families, so they had to sell everything. We had no tiffin work and we went back to the villages to work in the fields: two hundred *dabbawalas*. There were two hundred *dabbawalas* working for the cotton mills because we delivered on three shifts. Each *dabbawala* had about forty customers so there were two hundred with forty customers for each shift. Customers working both in the cotton mills and in the offices. There were office workers, managers and shopkeepers, but not those who did heavy work. Anyone who had a good salary ordered tiffin, middle-class people. Why would a low-earner order tiffin? Poor people brought food from home, in their hands or in their pockets. Nowadays we have customers among the people who work in the MIDC factories, the Maharashtra Industrial Development Corporation, but it is a very small industry. Nowadays, [the big] industries are located outside Bombay.

A member of the association also remembers:

> Tiffin work started with Raghunath Medge's father, who worked first of all at Bombay Churchgate station. When the British Raj was in power, then there was work at Churchgate. The people who worked at Girni had tiffin delivered to them. Bank workers were all Marathi and lived in a *cali*, which at that time was a building made of wood with long corridors and many single rooms. People went out early in the morning and tiffin was delivered later. Now Girni has closed. A man would deliver tiffin there, carrying it on

---

45   This data was provided by Raghunath Medge.

a tray on his head because there were no bicycles in those days, so we used the tray, and that was how the job started. As one man wasn't enough, one more was added, and since two weren't enough, a third was added. And now work has increased in the Girni, Lower Parel, Mahalakshmi, and Grant Road areas. And with time the network has grown. Any job where there are human relationships, stress and weariness grow and grow, and the work doesn't stop. It continues this way, it doesn't finish. The start was with the Girni workers. Now the cotton mills are closed in Mahalakshmi and Lower Parel, but that was where we started.

From the early 1900s until its slow decline at the end of the century, Bombay's economy was closely bound to the cotton industry. With the gradual shutdown of its main manufacturing companies, Bombay began its transformation process from a colonial industrial city to a metropolis where social and political dynamics began to play out after India's independence in 1947. Cotton mill profits were invested in the growth of an industry aimed at emancipating India from its ties with the United Kingdom and there was an attempt at a self-governing policy privileging the growth of the petrochemical, engineering and food industries. Growth lasted until the 1970s and brought with it increased employment in the city.[46] The changes also reflect in the history of mills that had shown a fairly balanced product quality profile until the 1950s.

Nonetheless, the city's post-1947 reconfiguration shows how the textile industry was forced to convert to fit into an economic system no longer bound by its (asymmetric) relations with British industry. Cotton mill production began to differentiate and work was decentralised outside of Bombay, with a progressive loss of bargaining power for the primarily Marathi-speaking workers. Product differentiation brought a demand for an unskilled workforce that was easily intimidated, which allowed for a maximisation of profits but at the expense of the safety, job stability and wage levels of the workers. This phase culminated with the great general strike of 1982–1983, when more than 250,000 workers gathered in the streets of Bombay.[47] The strike failed in its aim because mill owners used it as a pretext to close down unproductive sites, increasing the price of cotton clothing. The closure of the cotton industries also altered urban morphology, emptying the centre (which is geographically southernmost)

---

46 Sujata Patel and Jim Masselos (eds.), *Bombay and Mumbai: The City in Transition* (New Delhi: Oxford University Press, 2003).
47 See Herbert W. M. Van Wersch, *The Bombay Textile Strike, 1982–1983* (New Delhi: Oxford University Press, 1992).

of production facilities that were moved out into suburbs like Thane and Navi Mumbai.

The strike, which paralysed the cotton mills for almost eighteen months, had a significant impact on city life and on the work of the *dabbawalas*, since their main client base was forced to return to the villages after becoming unemployed. In Bombay's golden age of manufacturing, white-collar workers were apparently not the *dabbawalas'* main customer as they are today. An interesting degree dissertation on civil servants, for example, makes no mention of the use of a tiffin service as a preferred way of consuming the midday meal at work.[48] The NMTBSCT's secretary, Gangaram Talekar, explains the city's economic transformation and the consequent change of *dabbawala* clientele like this:

> In 1982 the cotton mills went on strike and at that time several *dabbawalas* were members, while others were employed. In those days *dabbawalas* took food from the *khanawals* and delivered it to customers in the cotton mills. There used to be a great many but they've closed now. The wives of the millworkers would cook and provide food. The word *khanawal* means someone who delivers tiffin, exactly like a *dabbawala*. But now they don't exist anymore. Lots of workers were our customers and we lost many after the strike. But we had no real problems at the association because it will continue to operate as it always has. Like a bus that has forty passengers on board, if some passengers get off, the bus carries on, what else can it do? We need to go on. We found new customers who arrived after the cotton mills closed and shopping centres opened in their place. The cotton mills have become business areas with small companies, factories, shops... They were there before, they're there today, and tomorrow they'll still be there.

... and the Mumbai middle classes

During the progressive decline of the textile district, other multinationals like Hindustan Unilever Limited (HUL) and Bata set up business in Bombay, facilitating the transition from an industrial to a service economy. In the last two decades of the twentieth century, Bombay recovered its role as a global city thanks to a number of key contemporary phenomena: the end of the bipolar geopolitical order; the emergence of new global governance frameworks; and the kind of social and cultural phenomena

---

48   R. N. Bhonsle, *Clerks in the City of Bombay* (unpublished MA thesis, University of Bombay, 1938).

that are considered typical of the postmodern condition. The city assumed a central role in the distribution of capital and information flows, and as a production platform for products and innovations in a post-industrial service industry economy.

The expansion of an economy linked to the service sector promoted the growth of new social classes, including a Marathi-speaking middle class, which became crucial for Mumbai's future.[49] Following the reorganisation of Indian states according to their linguistic identity in 1960, the state of Maharashtra was created by dividing the territory of the Bombay Presidency (which at the time included Gujarat). This was also the period when the Shiv Sena movement became established in Bombay, later evolving into a Hindu populist and nativist party. The Shiv Sena party suggested a programme of positive discrimination on an ethnic basis, seeking a voice for Marathi-speaking people. Although traditionally relegated to the lower rungs of the social ladder and jobs without status, the Marathas are demographically the largest ethnicity. They now began to demand a more important role in decision-making processes concerning municipal policies. The expression of this demand for economic and social representation evolved alongside the disappearance of classic forms of work organisation. There was above all an identity-type resistance, a possible vehicle of new forms of democracy but also of xenophobic outbreaks and religious fundamentalism.

In the mid-1990s, Bombay became Mumbai. The Shiv Sena party running the state of Maharashtra at the time proposed changing the city's name in line with a nation-wide process of altering names that were considered to be an expression of British influence. Mumbai derives from the name of the goddess Mumba, the divinity who is the patron of the Koli fishing tribe that originally inhabited the archipelago where the city developed.[50] This is how Medge describes it:

> 'Mumbai' means 'Mumba Ai'. Mumba is the name of the goddess and Ai means 'mother' in Marathi. There is the Temple of Mumba Devi. The goddess

---

49 See Patel and Masselos (2003).
50 Recent studies of the Koli goddesses of Mumbai have shown that the religious beliefs and practices of this community have changed since the late nineteenth century. Vicziany and Bapat suggested that Mumbadevi has been increasingly marginalized by the Kolis and has become absorbed into the practices of the Marathi and Gujarati communities of Mumbai. See Marika Vicziany and Jayant Bapat, "Mumbadevi and the Other Mother Goddess in Mumbai", *Modern Asian Studies*, 43 (2009), 511–41, DOI: 10.1017/S0026749X0700340X

protects the people here. Those who come to Bombay, who behave honestly, work honestly, will never find their bellies empty. This is what is believed, what is thought of the city of Bombay.

The change of name was popular with the Marathi-speaking people because it was already used in both Marathi and Gujarati, while those who spoke Hindi usually called the city Bambai. In this sense, the official name change emphasised the transition process from a colonial to an indigenous slant. Then, on 4 March 1996, the name of one of the most archetypal buildings of British Bombay was also changed when Victoria Railway Station became the Chattrapati Shivaji Terminus.[51]

The overwhelming reaction of the citizens to the city's change of name, was that they did not see this step as the semantic appropriation of a primary identity, but as the loss of what has been defined as Bombay's "proper name", the expression and metaphor of the diversity of India, the image of hope and modernity.[52] Here the lesson of historian Eric Hobsbawm comes to mind, suggesting that the invention of tradition was a strategy for the assertion of functional identity to achieve political ends particularly suited to the nationalist ideology.[53] Asserting the continuity of a suitably selected historical past allows practices of a ritual or symbol to be established in order to consolidate a social unit and corroborate its primal authenticity.[54] This was the meaning attributed to the change of name for many Indian cities: the desire to revive a primitive, "traditional", authentic, native vision of the place. This quest for authenticity actually achieves a converse effect of quite stale artificiality because, as Hobsbawm writes, "Where the old ways are still alive, traditions need neither be revived nor invented".[55] The name Bombay was an expression of the colonial era, but not an emblem of colonialism. It symbolised a multi-ethnic creole

---

51 The importance of the figure of the leader Shivaji in the city of Bombay and state of Maharashtra will be discussed in Chapter Two. "Shiv Sena", Shiva's Army (the reference is to Shivaji), has turned this historical figure into the cornerstone of nativist ideology (based on the concept of *bhumiputra*, "son of the soil"), which claims greater collective rights for the Maratha than those currently afforded to this population. The party reveals a strong Hindutva (Hindu fundamentalist and anti-Muslim) inspiration.

52 Thomas Blom Hansen, *Wages of Violence: Naming and Identity in Postcolonial Bombay* (Princeton: Princeton University Press, 2001).

53 See also Benedict Anderson, *Imagined Communities: Reflections on the Origin and Spread of Nationalism*, rev. edn. (London: Verso, 1991).

54 Eric Hobsbawm and Terence Ranger (eds.), *The Invention of Tradition* (Cambridge: Cambridge University Press, 1983), p. 8.

55 Ibid., p. 10.

cosmopolitanism, and here was an attempt to replace it with an identity of definite religious and ethnic connotation.

In the 1990s, Mumbai aligned with large global megalopolises, increasing the relevance of the tertiary sector, in particular financial services (the city is the home of the stock exchange), communications, IT, banking and the Bollywood film industry. The city was therefore driven to produce and market goods and services aimed at maintaining a lifestyle appropriate to a society in which a highly-developed tertiary sector plays an increasingly important role.[56] Despite everything, Mumbai continued to be distinctive for its huge variety of languages, religions, caste hierarchies, domestic rituals, festivities, forms of prayer and ways of dressing and cooking. All these aspects coexist in close proximity: sometimes they overlap; sometimes they are quite separate.[57]

It is precisely an attention to diversity that characterises Mumbaikar awareness, although cohabitation does require a continuous process of accommodation that is not always peaceful.[58] Mumbai culture does not appear to be the exclusive domain of a community, or a particular economic activity. The common denominator in this complex coexistence of overlapping hierarchies is a migrant identity: almost all Mumbaikars are immigrants or descend from immigrants. In the city there is a conviction that no one will be denied their chance of social mobility and self-realisation, a conviction that acts as a sort of horizontal organisation factor to alleviate the pressure to move up and advance in society. Moreover, the city continues to welcome and "handle" newcomers: small traders from South India; dairymen and farmers from Uttar Pradesh and Bihar; Hindu and Muslim taxi drivers; Goa Christian middlemen; wealthy Sindhi merchants from Karachi; and so on.[59]

The city's economic and social transformations have not decreased the *dabbawala* workload. The system has been able to adapt to the changes dictated by the constant flow of people who decide to live in the city for varying lengths of time, and proves that this delivery network can adapt perfectly to every new customer. According to one NMTBSCT *dabbawala*:

> With the closure of the cotton mills, we stopped delivering there. When businesses close, tiffin delivery stops. And if they close it's because there are

---

56   See Patel and Masselos (2003).
57   Sujata Patel and Alice Thoner (eds.), *Bombay: Mosaic of Modern Culture* (New Delhi: Oxford University Press, 1995), p. xiii.
58   Ibid. The violent clashes between Hindus and Muslims in 1992 to 1993 are a case in point, showing a progressive intolerance towards "foreigners".
59   Ibid., p. xix.

problems there, internal reasons. Now lots of companies have shut down in Bombay. They've moved. Some to Andheri, some to Malad, others to Kandivali, some of them outside Bombay. Their delivery service has been stopped. Like any company that has internal problems, right? Problems with payments or bonuses. The same goes for the cotton mills. Before the cotton mills closed, our network grew and grew in Bombay. What we did first for the cotton mills at Girgaon and Lower Parel, spread all over Bombay. From Andheri to Churchgate, we spread everywhere. That's why we haven't had problems after the closure of the cotton mills. Yes, the cotton mills have closed down, but that didn't cause us any problems.

As this eyewitness says, the changes to Mumbai's economic and industrial structure have altered the *dabbawalas'* customer profile and number to some extent. The data below show that despite the changes, there has been a constant increase in customers, dropping slightly only in the period that included the 1980s strike and Bombay's progressive social and economic reconfiguration.

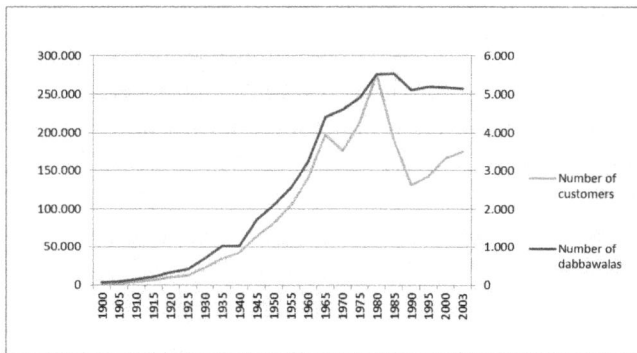

Figure 4. *Dabbawalas* and their custumers, 1900–2003.[60]

The data does not present an exact profile of the people using the *dabbawala* service today, but it does show that one of the main user bases is essentially that part of the Indian middle class with a fixed salary.

It is no easy task to define the Indian middle class, which is a highly diverse social group. The Indian National Council of Applied Economic Research defines as "middle class" those who earn between US $4,000 and

---

60   Data: Chandrasekhar, Ramasastry, *Dabbawallahs of Mumbai*, Richard Ivey School of Business, University of Western Ontario, 2004, available at http://beedie.sfu.ca/files/PDF/mba-new-student-portal/2011/MBA/Dabbawallahs_of_Mumbai_(A).pdf [accessed 28 October 2012], p. 17.

$21,000 a year, but these parameters can actually only be applied to six per cent of India's population. The economic proportion was then adjusted using a parameter of daily earnings that considered $5–$10 a day to be a middle-class income.[61] A recent study conducted by CNN-IBN and the *Hindustan Times*, on the other hand, suggests adopting a criterion based simply on the consumer profile, monitoring the ownership of possessions like cell phones, colour televisions and motor vehicles. This definition covers about twenty per cent of the Indian population. The difficulty in finding an exhaustive definition for the parameters of the middle class has led some scholars to wonder about the legitimacy of applying a typically western label to Indian society's intermediate income ranges. The concept of "middle class" is not merely statistical; it also embraces complex aspects like lifestyle, self-image, consumption profile, aspirations, etc. It might be more practical, as Rachel Dwyer says, to accept that the Indian middle class is very different from its western equivalent, since it includes civil servants, teachers, doctors, lawyers, white-collar workers, businessmen, but also film stars and military personnel.[62]

Some of these categories are upper-middle class, others are lower-middle, and they can be found in big cities that include Bombay and Calcutta as well as in smaller industrial and commercial towns.[63] The disparity between the large cosmopolitan cities and the rest of rural India shows how economic parameters are often inadequate for defining the identity of this slice of the population: what is needed for survival in a Bombay slum might offer a far more comfortable lifestyle in another location.[64] These inconsistencies are frequently perceived as a symptom of the social contradictions that undermine modern India and in part this can be seen in a dual definition of the middle class—traditional and modern, representing two distinct, complementary styles.[65] The traditional middle

---

61 Chakravarthi Ram-Prasad, "India's Middle Class Failure", *Prospect*, 30 September 2007.

62 Rachel Dwyer and Christopher Pinney, *Pleasure and the Nation: The History, Politics and Consumption of Public Culture in India* (New Delhi: Oxford University Press, 2001).

63 For an overview of the Indian middle class, see Pavan K. Varma, *The Great Indian Middle Class* (New Delhi: Penguin, 1998); Ruchira Ganguly-Scrase and Timothy J. Scrase, *Globalisation and the Middle Classes in India: The Social and Cultural Impact of Neoliberal Reforms* (London: Routledge, 2008); and Christoph Jaffrelot and Peter van der Veer, *Patterns of Middle Class Consumption in India and China* (New Delhi: Sage, 2008).

64 See William Mazzarella, "Middle Class", in the online encyclopaedia *South Asia Keywords*, ed. by Rachel Dwyer (2005), available at http://www.soas.ac.uk/southasianstudies/ keywords [accessed 26 June 2012].

65 Ibid., p. 5.

class can be traced back to the modernising drive promoted by Nehru after independence; it takes its cue from the old British elite's preferences and tends to ensure a certain continuity with the pre-independence administrative and linguistic model. The modern middle class, conversely, is attuned to what occurs in the planet's global metropoles, has a post-cosmopolitan colonial vision, and seeks the expression of a new Indian nationalism with a lifestyle moulded by specific consumer profiles.

This distinction is particularly important for research into the Mumbai meal delivery system. Although there is a huge amount of food available in the city, it is not considered like any other merchandise. If it were, customers would not use the *dabba* service: they would buy lunch anywhere. The decision to eat home-cooked food represents the continuity of a certain idea of tradition, suggesting that nourishment is actually what Arjun Appadurai calls "a powerful semiotic device", with tangible and intangible forms that are able to convey relationships with production and exchange.[66] The tiffin service means being able to eat home-cooked food at work and this simple act placates the city, reaffirming the specificities of each community within its boundaries.[67] On one hand material goods are expressed in a western, commercial language, while on the other, moral goods continue to speak the language of ethnic, caste and ritual purity traditions that cannot be understood if they are isolated from the person's cultural beliefs.

A small component of India's disparate middle-class population is represented by the businessmen who move frequently from one Indian city to another and also travel to major western capitals. These businessmen promote a new eating style hallmarked by the presence of foods with regional, caste and community influences typical of an ethnic origin, but revealing the influence of international western or Asian-style cuisine. Their wives do not necessarily have a link to their regional past, so they copy recipes from cookbooks written in English by Indians no longer living in India.[68] This category does not make frequent use of the *dabbawalas*, but

---

66   Arjun Appadurai, "Gastro-Politics in Hindu South Asia", *American Ethnologist*, 8 (1981), 494–511.

67   Madhugiri Saroja A. Rao, "Conservation and Change in Food Habits Among Migrants in India: A Study in Gastro-dynamics", in *Aspects in South Asian Food System: Food, Society and Culture*, ed. by Ravindra S. Khare and Madhugiri Saroja A. Rao (Durham, NC: Carolina Academic Press, 1985), pp. 121–40.

68   Arjun Appadurai, "How to Make a National Cuisine: Cookbooks in Contemporary India", *Comparative Studies in Society and History*, 30 (1988), 3–24.

does seem to embody a food style that is helpful in understanding part of the Mumbai population.

Today, the new horizon for the delivery service includes school children, staff in large shopping malls, new tertiary sector professionals and all those who want to eat home-cooked food with guaranteed hygiene standards in an increasingly polluted city.[69] Medge explains:

> People use our service because in Bombay pollution is really increasing: food, the use of bad oil, poor quality flour, all products you buy might be of poor quality. At home there's a wife, a mother, a sister, a daughter who cook the food themselves. The food is clean, you know what you're cooking… it's organic.[70] In Italy there's one type of food, but in Bombay it's not like that. There are many people with many cuisines: Punjabis eat one type of food, Marathis eat a different type, Gujaratis yet another type… and restaurant food may not appeal. It's not like the food cooked by a wife or a mother. There are Parsees, Christians, all kinds. In southern India they eat rice with fish, Punjabis eat rice with chickpeas, Gujaratis eat lentils, rice and *puri* [a type of fried bread]. In Bombay there are people of every community and they choose different items. The food prepared at home is better. The person who works in the office is a father, a husband, a son. Love is sent from home with the food, because it's been cooked by his mother, wife, daughter. Hygiene is guaranteed because once the wife has prepared food and put it in the container, the *dabbawala* never touches it. The tiffin contains a packet which is filled with food. The *dabbawala* wouldn't even have time to open the containers. The working day is from nine in the morning to midday, and they have to get moving before lunchtime to deliver, so they don't even have time to chat because they're so busy. There's pressure because they have to be on the train at these times. People trust the *dabbawala* so much that they put money, train tickets, a pen the husband may have forgotten, eyeglasses, medicines, all sorts in the tiffin. Sometimes, if there's been a quarrel, the wives may send a note that says 'sorry'. We pass on so many messages.

A job handed down from one generation to the next

As lifestyles come and go, tiffin delivery appears to be an element of continuity throughout Mumbai's food history. This continuity is the

---

69 There is no classification of new customers being served by the *dabbawalas*. My information is based on data acquired during field interviews. The schematic description applied should not, however, be seen as a definite demarcation of customer profiles and the timeframe included is simply to assist understanding.

70 In India the term "organic food" refers to products grown without the use of pesticides or chemical fertilisers, and inspired by the organic agriculture and animal liberation movements.

expression of the respect the *dabbawala* nurtures for the customer. The *dabbawala*-customer rapport is based on the understanding that good service guarantees survival for the worker and his family back in the village. As a result, the customers come to be seen as divinities because their patronage allows the service to thrive and continue. Medge puts it like this:

> We consider our customers as a god because they use our service. If customers didn't order our tiffin, our people, the *dabbawalas*, who are illiterate, wouldn't be given any sort of office job or similar work. Because they don't know how to read and write. The customers we take tiffin to are viewed as a god. With this work our members can help their family and the village, for education, for the sick. With this work, you earn money and serve God and the people. That is to say, faith is God. You find human satisfaction, personal satisfaction. This way, everyone is happy and Varkari Sampradaya is happy. You respect others and are given respect. You serve human beings. You do a good job. You donate food, so God will be pleased with you. This is the soul of Varkari Sampradaya. Through this teaching, all our *dabbawalas* put body and soul into their job. The *dabbawala* is happy, he has found the answer.

The most striking aspect of the framework Medge describes is the relationship of trust that has characterised the customer-*dabbawala* relationship over all these years, often lasting from one generation to the next. The work of delivering meals has its roots in a lengthy history of families serving one another, handing down the keys of the relationship to the next generation. Medge's story is typical, as he inherited first his father's job, then his position as president:

> I came to Bombay when I was seventeen or eighteen. As soon as I finished school. I took the twelfth-year exam in the village, then I came to Bombay to attend college, to study economics and geography. Then I left college. When my father died in 1980, I became the head of the family so I started work, and I had to leave college after the first three years. My father was president for twenty years, then Dipkant was president for eight... after him there was another president and the job's been mine since 1992. Actually, my father came to Bombay in 1940. It was my uncle who brought him here. My uncle came here in 1890 and he brought family and friends to Bombay because it was developing so fast and tiffin work just increased quickly. You need one man, and I mean a strong man, for every thirty tiffins. You have to load and unload your bicycle, deliver the tiffins, carry the tray on your head. My father brought the people he needed and that's how his wife and family came to Bombay also. My father ran this business from 1940 to 1980. My uncle came also. Two uncles and lots of other relatives came to Bombay because it developed so fast. The local railway got bigger. At the beginning,

the Bombay offices were in Fort Churchgate. So all the tiffins came from the city, to Fort Churchgate. Then the work left Fort Churchgate and got shifted to districts like MIDC, Virar, Borivali, Kandivali, Andheri, Kalyan, Thana, so *dabbawala* work spread there too. Now a tiffin can go from Andheri to Borivali, reach Virar, Churchgate, Thana, New Bombay, Vashi from Kurla. From Kurla to Maankhurd, from Maankhurd to New Bombay.

The continuity of the *dabbawala*-customer relationship over time does not depend only on the *dabbawala*: it also involves those customers who choose to use the tiffin delivery service that is a family tradition. Entire generations have eaten thanks to *dabbawalas* and have thus granted the service the characteristics of a rite of passage for entering a food system that is the expression of Bombay-Mumbai culture.[71] Being familiar with the delivery service means recognising the customs that regulate Mumbai's food supply and, in the words of Michael Herzfeld, acquiring "cultural intimacy" with an extended community, distinguishing the specific aspects that define shared sociability.[72]

The arrangement takes advantage of the convergence and correlation of different elements that Tullio Seppilli says can be defined in five main points:

a) a certain type of food;
b) structures and techniques that support the production and circulation/distribution processes, and the preparation practices for consumption;
c) institutional, organisational and behavioural methods applied to the consumption;
d) forms of technical training and conditioning and transmission of "models"; alongside
e) the cultural meanings and experiences of the psyche that refer directly or indirectly to food.[73]

---

71 Arnold Van Gennep, *Les Rites de Passage* (Paris: Picard, 1909). The author says that *communitas*, the rite of eating and drinking together, is a rite of aggregation, of union.
72 Michael Herzfeld, *Cultural Intimacy: Social Poetics in the Nation-State* (New York: Routledge, 1997).
73 I use the eating system definition found in Tullio Seppilli, "Per un'antropologia dell'alimentazione. Determinazioni, funzioni e significati psico-culturali della risposta sociale a un bisogno biologico", *La ricerca folklorica*, 30 (October 1994), 8–9.

The *dabbawalas* are consequently an expression of the Mumbai food system because in their context they bind together products, people, institutions and eating patterns through a network of relationships that merge into a recurring cycle. Awareness of being in this cycle, which returns regularly over the generations, gives the *dabbawalas* the self-assurance to manage their work. Lunch will always be a fundamental part of every working day, and this awareness allows the *dabbawalas* to see their future in positive terms. Talekar has this to say:

> Customers have confidence in the *dabbawalas* because they love their work and have no problems with tough work. Stealing from someone's wallet, stealing money: these are very bad things and in such things there's no satisfaction. You get no satisfaction. That comes from hard work, sweat. So they are happy at their work, no matter how tiring it is. We've had some customers for generations. For example, sometimes the father ordered tiffin, sometimes the brother, sometimes the son, so three generations have eaten thanks to us. Work is going well: there will always be people who eat, so there'll always be those who deliver their food. We delivered tiffin to the father, now to the husband, later we'll serve the children.

History itself is seen to be the *dabbawala*'s most loyal ally. People remember and recognise the *dabbawala*'s reputation to the point of authorising them to deliver their personal effects.[74] In more than a century of constant ferrying of food across Bombay, the faces of thousands of people who have used this system as a privileged eating system can be discerned. In the words of one NMTBSCT *dabbawala*:

> Customers really trust us and above all because they know the food is in it. For example, if you forget an important document at home, you know your *dabbawala* will arrive before one o'clock. So the *dabbawala* goes to your house on time and the person who gives him the food at home can put the document you forgot in with the food. We earn your trust and then we grow it. If the *dabbawalas* don't make mistakes in their work, then you'll trust them. Any company we work for trusts us if we don't make any mistakes. If we deliver for a year and there are no complaints about the tiffin delivery, then the customer can't suddenly complain, so they trust us. Bombs can explode and other things can happen, but trust in the *dabbawala* was already there, maybe for twenty years, maybe for forty years… it won't just go away.

---

74   Roberta Garruccio writes: "Reputation and trust are parallel aspects and that is precisely what makes the individual's reputation important for the group". See Roberta Garruccio, *Minoranze in affari. La formazione di un banchiere: Otto Joel* (Soveria Mannelli: Rubbettino, 2002), p. 39.

The distribution system also has sustainability characteristics not only in how it preserves the underpinning of a given society, but also for its interesting employment potential, offering opportunities to the poor and the illiterate. The *dabbawala*'s job is recognised as socially indispensable and requires significant skill in balancing food containers while travelling around the city at high speed on all kinds of transport. Besides the distribution system's very low costs and minimal environmental impact, it also gives some meaning to the lives of those who make it possible and it is perceived as a real "profession" that is handed down from one generation to the next.

A short Story: A Dabbawala Family

My father became a *dabbawala* twenty-five years ago; he is seventy now and he was born in a Konkan village called Raigar. He came to Bombay a few years after he married and when he was living here, he took another wife, so this was his second marriage. He has two daughters from his first wife and he has two sons and two daughters from his second wife. I am the oldest son and then there is my younger brother. He works with diamonds, he is a diamond quality controller. The office is near to our customers, where we deliver tiffins. My father came to Bombay on his own and he met my mother here. She also works as a *dabbawala*, her name is Lakshmibai, and we work together but her job is not so heavy: she doesn't carry such heavy weights as we do. I've been doing this job for four years. First I studied up to year twelve in Dadar, near Siddhi Vinayak Mandir. Then I did odd jobs… in Domino's Pizza and some electrician work, fixing cables. Then I decided to be a *dabbawala*: I decided I wanted to deliver tiffin. Now I'm thirty-eight, I live in Malad—Malad West—and I have a daughter. My mum helps look after our little girl when she finishes delivering tiffins, because my wife works. I like working here, I like it a lot. There is no stress. As long as you do your job, when you've finished you can go home. There is no pressure on *dabbawalas*. You do your job, you do it properly, deliver all the tiffins on time, that's the job.

---

2. *Dabbawala* Ethics in Transition

Varkari Sampradaya: faith and work

> They may be old or great, rich or poor, but they're all human beings. We have compassion for all human beings, we regard them with love. We distribute food to everyone we meet. We must help everyone. We protect our neighbours. If you have much, then give to the poor. God also gave life to the poor. Do not set them aside. God will give them something good. Everyone has feelings in their minds. Everyone should earn together and share. Eat together, putting things together, living together.
>
> — Raghunath Medge

The tiffin delivery network is not only supported by a complex logistics system (that will be explored further in the next chapter), but also by a special moral code. This code is the expression of the interrelationship between a specific manifestation of the Hindu faith—which can be traced back to the Varkari Sampradaya sect—and India's unique cultural philosophy. This sect places food at the centre of its philosophy, considering it to be a metaphor for life and its primary, material impulses and spiritual aspirations.

The Varkari Sampradaya ("the tradition of the masters") sect evolved in the wake of a drive for the renewal of the Hindu religious movement known as Bhakti, which preached pure devotion towards God as the way to salvation.[1] These movements developed from the fifth to the sixth century AD in Tamil Nadu, a state in the far south of India. Bhakti means "devoted love" or "loving union", and indicates a devotional practice not new to Hindu spirituality since it can be traced back through Vedic chants. In the tenth

---

1   Giovanni Filoramo (ed.), *Storia delle religioni IV: Religioni dell'India e dell'Estremo Oriente* (Rome: Laterza, 1996).

DOI: 10.11647/OBP.0031.02

century AD, Bhakti became a more widespread popular movement. Its main attraction and its revolutionary drive within the Hindu tradition came from the idea of a spiritual path open to all, without distinction of gender or caste. It ensured the believer would come into direct contact with God without the need for a go-between.[2] These movements clashed with Hindu orthodoxy, which decreed that the Brahmin castes were the masters of rites and go-betweens in the relationship between worshippers and God, ideas that consequently gave Brahmins an overwhelming social status and power.

Varkari Sampradaya is traditionally thought to have emerged around 1100–1120 AD, although this specific Bhakti movement was consolidated by Jnanadeva, whose work *Jnaneshvari* was actually written in about 1290.[3] This is a Marathi commentary on the *Bhagavad Gita* and is considered the Varkari bible. The text praises devotion to God and to gurus, whom the author says saved him from the corruption of worldly existence. It also celebrates the liberation obtained as a result of attaining mystical union with God; this union is the believer's ultimate aspiration, although it always remains outside of their grasp, given the immensity of the divine.[4] Jnanadeva was the first *sant*[5] of Varkari tradition which also includes Namadeva (c. 1270–1350), Tukarama (1568–1650), Ekanath (c. 1533–1599), as well as several women like Muktabai and Janabai.[6]

All these figures still inspire the spiritual beliefs of the Mumbai *dabbawalas*. This religious current in the Nutan Mumbai Tiffin Box Suppliers Charity Trust (NMTBSCT) is manifested in a strong sense of egalitarianism among its members, who come mostly from subordinate

---

2   For a detailed explanation of the manifestations of the Hindu gods Vishnu, Brahma and Shiva, see Alain Daniélou, *Mythes et Dieux de l'Inde. Le polythéisme hindou* (Paris: Editions du Rocher, 1992).

3   Felix Machado, *Jnaneshvari: Path to Liberation* (Mumbai: Somaiya, 1998).

4   Gavin Flood, *An Introduction to Hinduism* (Cambridge: Cambridge University, 1996).

5   *Sant* is the Hindi term used to define the religious mystic who acts as the spiritual representative within the community. It literally means "good man" and refers, in particular, to the figures of different castes born between the thirteenth and seventeenth centuries.

6   Mary Ford-Grabowsky, *Sacred Voices: Essential Women's Wisdom through the Ages* (New York: HarperCollins, 2002); Aliki Barnstone (ed.), *The Shambhala Anthology of Women's Spiritual Poetry* (Boston: Shambhala, 2002). Tukarama is probably the most venerated Maharashtra *sant*. Tukarama soteriology was based on the love of God but, unlike Jnanadeva's preachings, there was an evident distinction between God and humanity, as he believed that two separate identities are needed to develop a relationship of love. Moreover, Tukarama stated that meditation plays a fundamental role in devotion, because liberation can be obtained only when a person is seated in meditation, repeating the name of the Lord. See Flood (1996), p. 194.

castes or are even Outcastes or Dalits.[7] Namadeva, for instance, was a tailor, and Tukarama was of the Shudra caste. Recently however, several scholars have argued that in failing to make open criticism of the caste system, the Bhakti movement has involuntarily strengthened it. Furthermore, the disciples of the various Bhakti sects refused to marry in compliance with the membership rules of their own *varna* and were thus forced to marry amongst themselves. Inevitably this attitude actually created new castes.[8]

Sants preached not only on social and political issues, but also a doctrine of salvation that included devotion to the name of God, devotion to their own guru, and the importance of religious communion, of coming together in what is called literally "true community" (*satsang*).[9] The devotee must always be committed to the high moral values that are pivotal to the *sant*'s teachings because these values are not only a source of personal dignity, but also allow devotees to develop mutual respect. Even if a person leads a modest life or lives in outright poverty, their conduct must always be upright and pay service to God, expressing their Bhakti by serving other human beings. Helping others is the equivalent of an act of devotion to God.

Another unique aspect of Varkari Sampradaya is the importance attached to two female figures: in particular the mystical poet Muktabai, who was the sister and co-disciple of Jnanadeva; and Janabai, a servant of Namadeva who devoted verses to Vithoba, addressing the God as a female being named Vithabai. In Janabai's poetry, as in other works in this tradition, God can be both male and female, and may be addressed in the feminine, as one may address a mother. If the masculine is used, Vithoba is generally linked to Vishnu or the latter's *avatar* Krishna, and sometimes it is even linked to Shiva. Vithoba's cult defies sectarian division, as each year more than 6,000 Vishnu and Shiva devotees go on a pilgrimage to the Vithoba Temple in Pandharpur.[10]

---

7   It should be pointed out that Jnanadeva was a Brahmin excommunicated for failing to respect the orthodoxy required by his status.

8   Michelguglielmo Torri, *Storia dell'India* (Bari: Laterza, 2000), p. 139.

9   See Flood (1996), p. 193.

10   Ibid. In the Hindu context, the term God usually refers to Vishnu or Krishna (an *avatar* and personification of Vishnu), or sometimes to Shiva, although the worship of Vithoba goes beyond sectarian divisions. In the Trimurti, Vishnu the Immanent is the centripetal force that creates light; Brahma is the Immense Being, the orbiting force that creates space and time; and Shiva/Rudra is darkness, the centrifugal force, dispersing and destroying all that exists. See also Daniélou (1992).

In 1940, sociologist Irawati Karve took part in a Varkari pilgrimage to Pandharpur, writing a personal description of the experiences of the devout. At the time, the Samyukta Maharashtra movement was promoting the creation of a "united Maharashtra", whose political materialisation did not come about until two decades later.[11] The political commitment to this movement was alive during the pilgrimage and the great flux of Marathi speakers were united by an increasing recognition of their own regional identity.[12] For Karve, the pilgrimage was therefore not just a religious event but also represented a metaphor of Maharashtra, a way of giving it a collective definition.[13] This political dimension is echoed in the historical moment when the Varkari heritage meshed with that of the great Maratha leader Shivaji, who lived between 1627 and 1680. In the seventeenth century, the Varkari sect was the most important in Maharashtra and the *sant* Tukarama had a close relationship with Shivaji, archenemy of the Emperor Aurangzeb. It is very likely that in the struggle against the Mughal Empire many Varkaris fought in the ranks of its armies.

Varkari Sampradaya beliefs focus in no small way on the role that food plays in spiritual life. In the poems of the Maratha *sants* Tukarama, Ekanatha, Namadeva and Jnanadeva, food is present as a metaphor of the encounter with the divine. The worldly or spiritual meal has the task of teaching the eternal values of egalitarianism and brotherhood among the masses. The spiritual practice implies a collective experience of the divine banquet where all are welcome, no food is impure, everyone sits in a circle, nobody is untouchable and all are fed to satiety. The *sants* may have chosen the food metaphor because it is more understandable to devotees who cannot read or write.[14] Although the link between the ethics of the

---

11   The Samyukta Maharashtra Samiti was an organization of intellectuals and writers, founded in Pune in the 1950s by the leader Keshavrao Jedhe. It was formed to promote the creation of an independent state for Marathi speakers.

12   See Irawati Karve, "On the Road: A Maharashtrian Pilgrimage", in *The Experience of Hinduism: Essays on Religion in Maharashtra*, ed. by Eleanor Zelliot and Maxine Berntsen (New York: State University of New York Press, 1988), pp. 142–73.

13   Anne Feldhaus, *Connected Places: Region, Pilgrimage, and Geographical Imagination in India* (New York: Palgrave Macmillan, 2003).

14   Vidyut Aklujkar, "Sharing the Divine Feast: Evolution of Food Metaphor in Marathi Sant Poetry", in *The Eternal Food: Gastronomic Ideas and Experiences of Hindus and Buddhists*, ed. by R. S. Khare (Albany: State University of New York Press, 1992), pp. 95–116.

dabbawala association and the Marathi *sant* message is not always clear, it does seem that the food delivery work of the *dabbawalas* is inspired at least in part by traditional Varkari Sampradaya ideals.

Talking about himself, Raghunath Medge, president of the NMTBSCT, says:

> When I arrived, I was the only *dabbawala* with a degree and no one else had much of an education. Some had managed fourth year, some sixth, some first, some eighth. Then I understood that providing food really is the best gift of all. Giving food to someone is like serving God. Serving humanity is to serve God [it is serving God indirectly because God is in every person]. The personal intention is the supreme intention, which means that serving food will also bring earnings and receives God's blessings. Money is needed to keep your family. So serving food is the best gift. Serving people is like serving God. So the soul is satisfied. […] I earn 6,000 or 8,000 rupees per month, so I'm satisfied. I earn what I need to keep my family. We're happy with this because in our heart of hearts there is the best peace achieved by those who provide this service. So God will give me something in life. My soul is satisfied. Serving people is like serving God. Providing food is the greatest gift there is and it is an important aspect of Varkari Sampradaya. Everyone believes this so they work well. Work as a team, earn money, build your life: this is my idea. Work is worship. If I do my work well it's like practising *puja* [worship] of God.

It is difficult to say whether Medge's sentiment is shared by all the *dabbawalas*. Certainly the fact of forming a culturally homogeneous group allows members to identify with a shared religious and historical tradition. For example, the entire service takes a four-day break for pilgrimage to Pandharpur.[15]

---

15 There are two annual pilgrimages, which last 21 days, to Pandharpur. The first takes place during the Maratha month of *ashadi* (June/July); the second during *kartik* (November/December). The two pilgrimages culminate on the day of *ekadasi*, a day of austerity observed habitually by those who believe in *sanatan-dharma* or "Krishna consciousness". *Eka* means "one" and *dasi* is the feminine form of *dasa*, meaning "ten". *Ekadasi* is the eleventh day of the full or new moon, every month. On these special days devotees fast by abstaining from grains and legumes, and making a special effort to offer a service of devotion to Lord Krishna. On the day of fasting, if possible, physical effort should be avoided and the believers should dedicate themselves only to devotional services. It is believed that those who fast on this day obtain not only spiritual but also great physical benefits.

|| श्री कुलदेवता प्रसन्न ||
|| श्री ज्ञानेश्वर प्रसन्न ||

दूरध्वनी : २४२२ ९८२४
२३८६ ०७४२
२६८२ ९८९७

-: निवेदन :-

मुंबई जेवण डबे वाहतुक मंडळ यांजकडून तमाम ग्राहक बंधू आणि भगिनींना नम्र निवेदन करण्यात येत आहे की आमच्या गावी, तालुका मुळशी, मावळ, राजगुरूनगर, आंबेगाव, जुन्नर, अकोला, संगमनेर, जिल्हा पुणे या विभागातील गावो गावी कुलदेवताची यात्रा असल्यामुळे सर्व कामगार आणि मुकादम बंधू या भागातील असल्यामुळे त्यांना रजा देणे आवश्यक आहे. तेव्हा आपल्या जेवण डब्याची वाहतुक **शुक्रवार दि. ३०-३-२००७ ते बुधवारदि. ४-४-२००७** पर्यंत बंद राहिल वरील ५ दिवसाच्या रजे मध्ये महावीर जयंती ही १ दिवसाची रजा सरकारमान्य रजेव्यतिरिक्त आम्ही आपणाकडून ४ दिवसांची रजा घेत आहोत यांची ग्राहक बंधू भगिनींनी नोंद घ्यावी व आम्ही आपल्या होणाऱ्या गैरसोयीबद्दल दिलगीर आहोत. नेहमी प्रमाणे आम्ही **गुरुवार, दि. ५-४-२००७** या दिवसापासुन कामावर हजर राहून आपल्या जेवण डब्याची वाहतुक पूर्ववत चालू करू यांची ग्राहक बंधूंनी नोंद घ्यावी.

टीप : या रजेचा पगार कोणत्याही ग्राहक बंधूस कापुन दिला जाणार नाही.

नविन बर्ष सर्वांना सुख-समृद्धीचे व भरभराटीचे जावो.

आपले विश्वासू : **मुंबई जेवण डबे वाहतुक मंडळ,**
नवप्रभात चेंबर्स, ३रा माळा, रानडे रोड, दादर (प), मुंबई-४०० ०२८.

Tel. : 2422 9824 / 2386 0742

APPEL TO OUR PATRONS

All our patrons are hereby informed that the Annual village Deity festival of our Taluka Khed (Rajgurunagar), Mawal, Mulsi, Dist. Pune will be held from **Friday 30-3-2007 to Wednesday 4-4-2007** (both day including). Since all of us come from the same village we all are required to attend the festival, Tiffin Carriers Service therefore will remain suspended for a 5 days from Thursday 5-4-2007 inconvenience caused to our Patrons is very much regretted resume the service from as usual. It should be noted that there shall be no deduction in wages for 1. day during which our service is suspended During this period Mahavir Jayanti is already declared Public Holiday. For more information contact Mukadam on Tel.

Thanking you.

Yours Faithfully,
Mumbai Tiffin Carriers Association.

Figure 5. Mumbai. This flyer informs customers that the service will be suspended for four days for the annual festival of the *dabbawala* villages of origin and one day for the national Mahavir Jayanti holiday, which falls between late March and early April of the Gregorian calendar and celebrates the birth of Mahavira, the spiritual teacher of Jainism. By kind permission of Raghunath Medge.

A tangible sign of the shared religious faith of the *dabbawalas* can also be seen in the *dharamshalas*, which are stopovers close to temples where pilgrims can stop to rest. *Dharamshalas* were erected in Bhimashankar in 1930, in Alandi in 1950, in Jejori in 1984, and in Pandharpur in 2000, complying with the wishes of Madhu Havji Bacche, founder of the NMTBSCT. He had initiated construction of the first two *dharamshalas* and over the years various *dabbawala* groups contributed to the completion of others via a voluntary donation system. Medge tells the story:

> Mr Bacche was Varkari and before coming to Bombay he arranged for *dharamshalas* to be built. The first was built in 1930 in Bhimashankar, our home village. There is the Jyotilinga Shankarji of Bhagavan, which we call Jyotilinga. We call his Shankarji temple Jyotilinga. But the *dabbawalas* have built four *dharamshalas*. One is Bhimashankar, the second is Alandi, the third is Jejori, and the fourth is Pandharpur. Tukaram, Jnaneshvar and Ekanath have written about Rambal Krishna [stories related to Krishna's childhood], as well as several poems about food. Also more generally about life, about *samskar* [the religious ritual that marks the main moments of Hindu life], about responsibilities people have. How to be devoted to God. How to respect *izzat* [family honour and prestige], the elders in the family. How to provide for your parents, worship God. Everything has been written in them. Nivritti is master of Jnanadeva, Mukta which is the abbreviation of Muktabai, Sopa who is the brother of Jnanadeva and Muktabai.[16]

A Short Story: The Dharamshala Caretaker

I was born in Bombay and I started as a *dabbawala* when I was twelve. My dad was a *dabbawala* and he worked with Medge's father. I never met Bacche, I only knew that he was an important person. He was the one who organised the *dabbawala* association, turned it into a working group. When they were in Bombay, Bacche, Medge's father, and my dad stayed together. Living on the street. Bacche also had the idea of building the first *dharamshala* at Bhimashank. They asked for donations, one rupee, two rupees, five rupees. You can still see all the names. The money was collected over a couple of years: the *mukadams* gave most money because they ran the line and the *dabbawala* were employees. After he had collected donations, Bacche bought land and built the *dharamshala*. Over the years, we built the shops alongside. Here at Alandi there are

16   My thanks to Pinuccia Caracchi, Professor of Hindi Language and Literature at the Faculty of Foreign Languages and Literature, Turin University (Pinuccia Caracchi, personal correspondence).

lots of *dharamshalas*: the immigrants pay for them to be built, for example the fishmongers have one and so do the greengrocers. But we were the first. When I couldn't work as a *dabbawala* anymore, after I had three bicycle accidents, Medge asked me to come and be the caretaker here, at the Alandi *dharamshala*. I live here, where I have a room with my wife.

Another aspect of *dabbawalas'* work that seems to be in line with the Varkari Sampradaya worship ethic is a belief that the human being is a go-between with God. There is a perception that food delivery constitutes an act of religious devotion that reveals the absence of discrimination toward others. Just as the Varkari Sampradaya devotees consider life to be a pilgrimage, the *dabbawalas* are constantly on the move for their work and for their faith. As one NMTBTC *dabbawalas* describes:

> Our tradition is really to see work as *puja*. First of all, in our work there's no discrimination. All of us are Hindus but we bring tiffin to Muslims, Christians, Gujaratis, and any strangers. But we don't discriminate a Hindu or a Muslim, or any other caste. We certainly don't. In the same way we lovingly serve food to a Hindu, we also serve others, like Muslims or Christians [...] This aspect of our work has never changed and there has never been a reaction like: "Don't touch a Muslim's *dabba* because he's not vegetarian". If we deliver food to a Hindu, we also deliver to Muslims. Our distribution network is still the same and doesn't change. Pick up tiffin, deliver tiffin: if we make a mistake and the customer gets angry, we accept it. We never say: "Brother, that's another *jati's!*" Anyone can do a job for money, but tiffin work is different. You can't discriminate or make caste and religious distinctions. Do your job, that's it. For example, at Marine Line there's a Muslim area and anyone who goes there is afraid and removes their *topi* [the cap worn by Hindus]. Hindu workers take off their hats when they go to a Muslim area. But we're not worried because we're doing our job and not discriminating anyone. Anyone who discriminates is afraid. We're not frightened. *Dabbawalas* work there every day because we have to earn money, but earning can be in different ways. This way, there is hard work. In human relationships, delivering food to another is a kind of job that means you cannot discriminate.

The origins of a lineage

> The warrior prince Shivaji is our ancestor, the father of the father of our paternal grandfather; the whole family descends from him. Chattrapati Shivaji was king; Chattrapati Shivaji was an emperor. We were first his soldiers but that is now in the past. Now we don't have soldiers any more. Now there is the government. We have to earn money and to do that we

have to work. Now we pay service to sadhus [ascetics] and sants. [...] This is strength. This is devotion. Strength is needed for all work. We learned this from him. To do business, kindness is required. We learned that from them. We learned kindness from the sants and courage from the emperor. We believe in the same Sampradaya as the pilgrims. Our families believe that providing food is punya, a worthy action that brings religious virtue: work is worship. Serving food is considered a worthy action.

—Raghunath Medge

The *dabbawalas* define themselves as descendants of the warrior prince Shivaji Bhonsle. He is considered by his followers to be the founder of the Maratha nation because of his relentless struggle against the hegemony of the Mughal Empire. The Emperor Aurangzeb, who contemptuously called him the "Deccan mountain rat" never managed to defeat the tireless rebel. The Emperor's objective was the gradual transformation of the Empire into a Muslim State, which was implemented—although never completely and never successfully—through the introduction of Islamic law and the elimination of a series of symbols that supported a secular state. The response was a Maratha battle against Muslims. Through guerrilla warfare Marathas avoided fighting out in the open, but destroyed enemy lines of communication and assaulted isolated detachments. This approach required a large number of soldiers, many of whom were recruited amongst Maharashtra farmers, who are probably the ancestors of today's *dabbawalas*. Shivaji's profound knowledge of the territory allowed him to achieve military success, which brought him a reputation as one of the fathers of modern guerrilla warfare. In 1674, during a traditional Hindu ceremony, he was crowned Chattrapati or "Lord of the Universe" by Ramdas, a *sant* of the Varkari tradition. Contemporary Indians consider Shivaji important for his contribution to the forging of a proud Hindu nationalist spirit in his people.[17] The *dabbawalas* see themselves as bonded to Shivaji by a shared Hindu faith, a fierce sense of independence from any domination, and patriotism towards the State of Maharashtra.

Shivaji is a pivotal figure in Maharashtrian beliefs, fundamental to the understanding of events that developed the political and social scene of recent decades. The mythology of Shivaji is crucial to understand the symbolic reconstruction that underpins Indian systems of political rhetoric, in particular those of the Marathas. Sometimes the interpretations

---

17   Stanley Wolpert, *A New History of India* (Oxford: Oxford University Press, 1977).

of Shivaji's contribution differ between Western and Indian scholars. Without detracting from the importance of the Indian studies, they are often ideologically oriented within the Hindu nationalist movement, which reconstructs Shivaji's heroics and the Maratha movement using not entirely reliable historical sources. For example, in a work by the judge and reformer Mahadev Govind Ranade, Maharashtra's cultural unity already had its own common language with an important literary tradition, which then evolved into modern Marathi. According to Ranade, this process of linguistic and symbolic amalgamation may also have led to the development of shared social structures and moral codes.

In this perspective, Shivaji's role was similar to an enzyme in a catalytic process that has already started, bringing together all Marathi-speaking people under one banner and instilling in them a stronger sense of cohesion and community. Ranade also attaches great importance to the Maharashtrian Bhakti movement, which is to say the Varkari Sampradaya that promoted a society without castes.[18] This, as well as other historical theories that emphasise Shivaji's non-Brahmin dimension, highlight how Maharashtra's political and religious history comprises a complex set of icons, symbols, and proto-ideological ideas which twentieth-century local Indian politics drew upon to develop an independent ideological scheme.

Evidence that the anti-caste drive and religious renewal that had crossed the region from at least the twelfth century became more powerful under Shivaji is also found in the reconstruction of a dispute between Shivaji and local Brahmins, who had denied Kshatriya status to the Kunbi farming caste that later evolved into the Maratha warrior caste.[19] Shivaji opposed this decision by handing out privileges and powers to Kunbis who distinguished themselves by their service (mostly military) to the monarch. This challenge to the Brahmin establishment is always an inspirational presence in the Bhakti movement. The story also gives a new symbolic significance to the pre-Aryan divinities: for instance, legend has it that the goddess Bhavani gave Shivaji her invincible sword.[20]

---

18   Interpretations were reconstructed thanks to Thomas Blom Hansen, *Wages of Violence: Naming and Identity in Postcolonial Bombay* (Princeton: Princeton University Press, 2001), pp. 26–27. For further information, see also Mahadeo Govind Ranade, *The Rise of Maratha Power* (Bombay: Bombay University Press, 1961). The author elaborated the work of Govind Sakharam Sardesai, *New History of Marathas* (Bombay: Bombay University Press, 1946).

19   Warrior ideology was one of the Shiv Sena's key tools to penetrate the Maratha Hindu imagination.

20   See Hansen (2001).

Further proof of Shivaji's revolutionary potential can also be found in the post-Independence appearance of readings inspired by Ghandian and Marxist thought, celebrating Shivaji as the enlightened ruler who abolished forced labour and became the symbol of the battle to end the caste system.[21]

Notwithstanding the merits of their sources, these various historiographical interpretations stress that Shivaji is generally considered to be the father of Maharashtrian nationalism, a powerful archetype who can bring new life to Hindu identity. Shivaji's appeal was an important factor in the political debate surrounding the formation of the State of Maharashtra and the development of post-colonial Mumbai's urban culture. It appears that the ethical canons chosen by the *dabbawalas* are also taken from this legendary ancestor, including gender equality, non-discrimination, Hindu religious beliefs, and the idea of work as a source of strength and liberation from poverty. A retired *dabbawala* and son of the NMTBSCT founder says:

> At the beginning there was only one group of *dabbawalas*, then other groups formed. Then the groups joined together to create the association. That's the thing about *dabbawala* work: we can get jobs for everyone if we're a group, and that allows us to deliver tiffin. If I work alone, I won't manage. If you serve people, God blesses you, and serving people is like serving God. The spirit is content if a job is done well, if it is performed properly. This is the Varkari Sampradaya way: live correctly, earn correctly, work correctly. Do not take work from others. Do not earn illegally. This is our Varkari Sampradaya law. We Varkaris do not see any differences to do with *jati*.

Caste and descent

Caste is a complex concept and should always be approached with caution and sensitivity given its social, economic, ethnic and religious implications. Many Indian intellectuals are uneasy when they see their society constantly described and interpreted through caste. There are primarily two reasons for this: the first is linked to the awareness that the analysis of India's caste system is largely a product of western Orientalism;[22] the other lies in

---

21 See Meena Menon and Neera Adarkar, *One Hundred Years One Hundred Voices: The Millworkers of Girangaon: An Oral History* (New Delhi: Seagull Books, 2004).

22 One European author who looked at the caste concept was French scholar Louis Dumont, certainly from a privileged standpoint: see Louis Dumont, *Homo hierarchicus. Le système des castes et ses implications* (Paris: Gallimard, 1966). Nonetheless, a complex debate on the subject has also taken place in India. See Dipankar Gupta (ed.), *Social Stratification* (New Delhi: Oxford University Press, 1991); and R. S. Khare (ed.), *Caste, Hierachy*

the fact that this analysis seems to reduce the complexity of India's (and its diaspora's) social and economic development to archaic, unchanging categories, without adequate consideration of the massive transformations that modernisation processes and migration dynamics have triggered in India. Rules regarding contamination caused by contact with individuals considered inferior by birth or for frequenting specific places (like hospitals) are being increasingly ignored. The connection between caste and professional specialisations, especially in urban contexts, has also lessened considerably.[23]

According to Ronald Inden, for far too long western scholars have described Indian society as essentially condemned to backwardness for a number of reasons, with a strong emphasis on the caste system.[24] The formal codifications of the system are proposed as a tool for interpreting the present without verifying what the reality has actually become. For Inden, this ahistorical and ingrained approach to Indian civilisation and society constitutes the main weakness of many Indological perspectives of the nineteenth and twentieth centuries. More recently in India, and particularly in the state of Maharashtra, caste rhetoric has been put to controversial political use, especially by the neo-Hindu movement of which the Bharatiya Janata Party is the leading political player. Looking for an alleged racial authenticity, the party used Mumbai as its stage for economic and social demands increasingly and explicitly based on linguistic and caste affiliation.[25]

To avoid these methodological and ideological pitfalls, it would seem best to introduce a historical perspective for analysing the relevance of caste in relation to the social organisation of the *dabbawalas*.[26] The main stages of this interpretive path are based on an introduction to the concept of caste, an analysis of the bond that links the formation of the Maratha caste with the

---

and Individualism: Indian Critiques of Luis Dumont's Contributions (New Delhi: Oxford University Press, 2006).

23   Mysore Narasimhachar Srinivas and M. N. Panini, "Casta", in *Enciclopedia delle scienze sociali*, ed. by G. Bedecchi, vol. 1 (Rome: Treccani, 1991).

24   Ronald Inden, *Imagining India* (Oxford: Blackwell, 1992).

25   For an overview of Mumbai's political and social situation I recommend the meticulous historical reconstruction in Jim Masselos, *The City in Action: Bombay Struggles for Power* (New Delhi: Oxford University Press, 2007).

26   I am aware that this perspective is based on sources selected by these same Orientalists, attaching a stereotypical reading of Indian society and culture, and is therefore liable to incur the same generalisations and inaccuracies. Hence the effort to use their reconstructions with due caution, and to bear in mind the scientific debate that arose around those issues.

historical figure of the warrior Shivaji and, lastly, the contextualisation of the development of caste dynamics in Bombay's industrial and commercial history. A detailed discussion of the political and social life of the city in recent decades does not fall within the remit of this book but, in order to understand the collective *dabbawala* experience, it is necessary to give a brief overview of the particular cultural-political sphere of which it is part.

A brief introduction to the caste concept

The main traditional sources of Hindu doctrine, the *Purusha-sukta* hymn in the *Rigveda* (late second and early first millennium BC) and the *Manu Smrti* (the *Laws of Manu*, first millennium BC), indicate that in the caste system the social order is governed by a hierarchy evolved on the basis of ritual purity. The constituent elements of this social order are the categories (identified by means of their *varna*, or symbolic "colour") attributed with the three fundamental functions of Indo-Aryan society from the time of its origins. The Brahmins, or priests, have the colour white; the Kshatriya or princes and noble warriors are designated with the colour red; the *vaishya* or the people (farmers and traders) are symbolically yellow. These three categories (called *varna arya* or "noble colours"), found in all the different populations that speak Indo-European languages, then became four, with the addition of the *shudra*, given the symbolic colour black and representing the mass of non-nobles (*varna an-arya*) referred to by the generic name of *dasa* ("servants").

The *Laws of Manu* entrust the *dasa* with the task of "serving" the three noble categories also called *dvija* (which means regenerated or born twice, as their official entrance into society occurs through a rite of initiation that marks a second birth, conferring them with a further positive quality). The different groups of people and, in the broadest sense, the different forms of existence acquired at birth within the *varna*, are called *jati* (a word derived from the Sanskrit root *jan,* meaning "to generate", a term that incorporates functional and hereditary caste aspects). The actual word "caste" is a Portuguese translation of the term *jati*, which missionaries rendered with a term that meant "pure" in their language.[27] The first three *varnas*, those of the *dvijas*, at least theoretically constitute an equal number of *jati* whose internal subdivisions are usually regarded as subcastes or *upajatis*, and

---

27 The word *jat*, which shares the root of the Latin words *gens* and *genus*, indicates a concept akin to the English word "kind" and suggests a range of meanings similar to those expressed by the words family, ancestry, lineage, kinship, rank and race.

differ amongst themselves in accordance with family, type of employment, or origin, while a huge number of *jatis* converge in the *shudra varnas*, distinguished mainly on the basis of employment or service rendered to the dominant caste.[28]

Critical interpretations of Hindu ideology and the caste system tend to fall into two different, but in some respects complementary, approaches. The first, preferred above all by Max Weber, fuses the institutionalisation of inequality with the concepts of *dharma, samsara* and *karma*.[29] *Dharma*, which is "moral order", is the ethical imperative that urges the individual to comply with the divine laws of conduct associated with each caste. If the person performs this task adequately, their *karma*, or "a present action able to influence their rebirth positively or negatively", will be positive and after their death will allow their spirit to flow back into the *samsara*, or "the eternal cycle of births and deaths", with the prospect of reincarnation in a higher caste. Otherwise, they will be reborn into a lower caste or even inferior life forms like animals, plants, etc. The relationship that links the various castes is thus arranged to comply with principles of complementarity and consequently everyone should be eager to maintain a social order that conforms to a moral order (*dharma*).[30] The second approach is that proposed by Louis Dumont, which places the emphasis on the concepts of purity and contamination, as well as on the strict separation of religion from politics and economics.[31] Priests are afforded a higher status, beyond considerations of economic or political power, and the caste system should respect compliance of the human sphere with *dharma* laws, therefore contributing to balance and harmony.

Both approaches have generated extensive criticisms and theoretical developments, and debate continues on the ideological, ethical-religious, political, economic, social and cultural foundations of the caste system.[32] Stefano Piano writes that historically the Indian caste system has assumed the characteristics of a closed social group, defined almost exclusively

---

28   This simplified explanation of the constituent elements of the caste system draws broadly from Stefano Piano, "Lo hinduismo II. La prassi religiosa", in *Hinduismo*, ed. by Giovanni Filoramo (Rome: Laterza, 2002), pp. 171-246.

29   Max Weber, "Economic Ethics of the World Religions: New Perspectives. Part Two. The Sociology of Hinduism and Buddhism", in *The Sociology of Religion*, rev. ed. (London: Methuen, 1965).

30   Adrian C. Mayer, *Caste and Kinship in Central India* (Berkeley: University of California Press, 1960).

31   See Dumont (1966).

32   See Gupta (1991); see also Khare (2006).

by birth. It includes a number of families and is often, but not always, associated with employment. It is quite significantly characterised by ethnicity and religious or geographical origin and intermarriage. The behaviour of its members is influenced by precise dietary and shared eating rules.[33] In more general sociological terms, a caste can be defined as an ascriptive aggregation by right of birth, rigidly superordinated or subordinated to other aggregations of the same type, within a social stratification system in which individuals are not permitted any vertical mobility. David Mamo writes that this interpretation of castes indicates they are distinct social and cultural entities. As such, they tend to create internal subcultures as a consequence of the intensity and quality of communication within the group, compared to that of the group with others. In fact, such communication is fostered and preferred as it is functional to the strengthening of a sense of caste identity.[34]

Caste and industrialisation in Bombay

As discussed in Chapter One, caste, kinship, and rural-urban relations or rural connections were fundamental to Bombay's social organisation, so much so that some historians and anthropologists have defined Indian cities as "an urban landscape composed of rural institutions".[35] It is difficult to estimate the exact weight of the caste system either in old Bombay or in modern-day Mumbai, because there has been a significant propensity to believe that castes are not an appropriate social organisation for an industrial urban context.

Davis Kingsley put forward the hypothesis that, with the rapid increase of industrialisation, the caste system would disappear.[36] On the other hand, historian Rajnarayan Chandavarkar posited that the current notion of caste would be reformulated incrementally to reflect the growing importance of cities in Indian society. Morris D. Morris, however, argues that the alleged disappearance of castes in the cities is mainly due to the difficulty of

---

33 Piano (2002), p. 172.
34 D. Mamo, "Casta", in *Nuovo dizionario di sociologia*, ed. by Franco Demarchi, Aldo Ellena, Bernardo Cattarinussi (Milan: San Paolo, 1987), pp. 336–40.
35 Rajnarayan Chandavarkar, *The Origins of Industrial Capitalism in India: Business Strategies and the Working Classes in Bombay, 1900–1940* (Cambridge: Cambridge University Press, 1994), p. 219.
36 Kingsley Davis, *The Population of India and Pakistan* (Princeton: Princeton University, 1951), p. 176.

finding reliable historical sources to document their existence, and that the role of caste dynamics in Bombay's early industrial labour market has been underplayed.[37] Some scholars have also suggested that caste dynamics exist mainly within the private social sphere and are actually disappearing from the public world.

Bombay's industrial and social history has shown, instead, that it has two contexts: one public and one private. They are always closely interconnected and interdependent because the social organisation of urban workplaces is linked to neighbourhood and family relationships, as well as regional origins.[38] There is little data on the caste composition of Bombay workers, but according to a 1940s survey commissioned by the Bombay Mill Owners Association (BMOA), which referred to the workforce in nineteen factories, most seemed to be Marathas and Kunbi Hindus (approximately 51.8%); Bhayas make up 13.8%; Untouchables accounted for 11.9% and the remaining 5.2% were Muslims.[39]

The difficulty in defining a worker caste profile lies in the fact that factory register entries are generic, and indicate only religion, *jati* and place of origin; moreover they lack uniform and mutually exclusive criteria.[40] The survey mentioned by Morris noted a relatively low percentage of Untouchables, but that seems to have grown subsequently.[41] It is likely that during the launch of the first cotton mills there were few Untouchables in the city since other workers were reluctant to be in the vicinity of Dalits. A document issued by a United Spinning and Weaving Mills manager in 1874 prohibited Untouchables from working in these factories. In the early twentieth century, the number of Untouchables rose significantly and stabilised in the 1920s and 1930s, during industrial strikes. The Dalits' greater vulnerability meant they could be blackmailed and were often recruited as strike-breakers. During the 1929 general strike, for example, the prominent political Dalit leader Ambedkar strove to provide the workforce needed to continue cotton production.[42] Ambedkar became an icon of the Dalit struggle for emancipation and

---

37   Morris D. Morris, "The Emergence of an Industrial Labour Force in India", in *Social Stratification*, ed. by Dipankar Gupta (New Delhi: Oxford University Press, 1991), pp. 231–47 (p. 240).

38   For further information, see also David West Rudner, *Caste and Capitalism in Colonial India: The Nattukottai Chettiars* (Berkeley: University of California Press, 1994).

39   Morris (1991), p. 240.

40   Chandavarkar (1994), p. 220.

41   Morris (1991), p. 245.

42   Ibid.

he used this to consolidate the presence of the Untouchables in the industrial workforce, convinced that this might help to reinforce their political capacity and social position.

Importantly, the urban context has partly changed the caste concept by breaking down distinctions based on intermarriage, eating habits, and shared linguistic, regional or religious traits. Bombay's biggest caste, the Marathas, is actually deemed to be the expression of broader caste units, allied to improve their chances on the city's labour market.[43] Uniting in associations governed by the most influential members, the so-called *dadas*,[44] these groups of people of disparate caste origin are nonetheless able to identify with a wider Maratha denomination. These multi-caste conglomerates often organised whip-rounds to raise funds, and set up a subscription system to ensure welfare and job mediation services for their affiliates.[45] Maharashtra society, which is characterised by a certain linguistic and cultural homogeneity that probably facilitated this process, saw numerous *jati* farmers join the Maratha caste. Through these new affiliations, Rajnarayan Chandavarkar suggests that "caste identities came to be expressed in caste associations which could operate throughout the city".[46] Although they could all be traced back to the same Maratha designation, these "caste associations" allow their members to express their experience of the city on the basis of the various affiliations that made most sense: the village of origin: a shared language (dialect), the district of residence in Bombay, level of education, etc. All of these are transversal affiliations connected to the person's past that allow them to transcend their *jati* or *upajati* origins.

The NMTBSCT has all the features of these caste associations. In addition to the prominence given by the internal hierarchies to the Maratha identity and its most important icons (starting with Chattrapati Shivaji), there were also the common rural origins of many affiliates and a shared

---

43 See Chandavarkar (1994).
44 *Dada* literally means "grandfather" in Hindi but is used in a colloquial manner to indicate the older siblings or figures from a criminal subculture. *"Dada* culture" has a long tradition in Bombay's industrial history as a popular model of authority and power. *Dadas* often come from humble backgrounds and thanks to strategic relationships can become leaders of associations and bands, and middlemen for finding work. The term *dada* in this context indicates the exercising of political power that evokes the virile image of a man able to trigger actions in his social context. See Hansen (2001), p. 72.
45 See Chandavarkar (1994).
46 Ibid., p. 234.

language base. Gangaram Talekar, the NMTBSCT secretary, describes the *dabbawala* affiliation:

> I'm from the Pune district and I came to Mumbai to do tiffin work. The name of my village is Rajgurunagar and it's near Pune. We're from there [...] Rajgurunagar, Ambegaon, Maval, Munshi, Akola, Sangamner. All of us *dabbawalas* come from there and we've been doing this work for a few years. I'm the third generation of tiffin workers: at the beginning my father did it, and his brother-in-law worked with him. Then in 1956 I started too. My family and Medge's worked together, the two families are like one family. We're very close... like brothers.[47] We're all one family [points to a boy who works as a *dabbawala*] and it's as if he was the son of an uncle or brother. That's how it is with *dabbawalas*, everyone is related to someone else.

One thing that seems to reinforce the association's identity matrix (although it should be put into the context of a relational mode that is common and widespread throughout India, and which is part of a consolidated code of conduct) is the habit of calling their colleagues with nicknames that refer to the family, and which are assigned according to the age of the person in question. *Dabbawalas* will call an older person *dada* (paternal grandfather) or *kaka* (younger paternal uncle, in other words the father's younger brother). Peers may call each other *bhai, bhau, bhaiya* or "brother". One of four female *dabbawalas* in the association is called *mami* or "aunt" (mother's brother's wife). Medge explains:

> There are also women *dabbawalas* who do light work. For example, if the husband is ill, the woman goes to work in the fields. If someone is injured, is in hospital, isn't feeling well, the woman has to help out. Women are given light tasks in the work group. Take Lakshmibai: she works in the Santacruz station and lives in Kandivali. We *dabbawalas* call her 'mami'. We often call each other with the names of family members... *dada, bhai, bhau*. For instance, in his group Choudhary is known as Mukadam, Dada Mukadam, Choudhary Mukadam. As happens in a family.

Just as labour recruitment in Bombay's large cotton mills during the first decades of the twentieth century occurred through complex interwoven links among castes, family, and neighbourhood, today's channelling of

---

47   Psychoanalyst Sudhir Kakar believes that the bond between brothers is the basis of the extended Indian family. In the ideal extended family, the brothers remain together after marriage, bringing their wives into the family circle. The concept of brotherhood includes fraternal loyalty, which resounds in the economic, social and ritual facets of the extended group. See Sudhir Kakar and Katharina Kakar, *The Indians: Portrait of a People* (New Delhi: Penguin, 2007), p. 9.

the labour force in Mumbai's tertiary economy occurs in the same way, especially in contexts where informal economy is prevalent. If the caste associations, of which the NMTBSCT is an example, have been able to act as engines of empowerment for their members, enhancing their contractual position (and, of course, that of the associations themselves), it is largely due to the strong symbolic and affective intensity of the relationships among their members.

The key to the transmission and promotion of these interdependencies has always been the family, which continues to act as a mediator between the private and the public spheres, even in the metropolis. On one hand, the family maintains its significance as the matrix for organising migration plans, marriages between the same *jatis*, or professional careers. On the other, it stands as a relationship model in an urban context of social relations shaped by supra-urban and supra-national economic, social and political forces. The metropolitan arena is a place where various caste associations live in close contact and never achieve cultural equality, but simply create a dense fabric of transversal relations that find their identity in the interrelation within these associations.

The question of the *dabbawala* "caste"

> Our caste is Maratha, from Maharashtra. It's Hindu. Shivaji was a Maratha Hindu.
>
> —Raghunath Medge

In this composite cultural and social order, the *dabbawalas* see themselves as a "Maratha caste", part of the Kshatriya *varna* of warriors and fighters. As already seen, however, most of the *dabbawalas* originate from the western part of the state of Maharashtra, from small rural villages. Although they are not so far away from the city, their areas of origin have a completely different landscape to that of the city, dominated by cultivated fields and hills covered with sparse bushes. The economy of the home villages is predominantly agricultural and most work derives from this sector: agriculture and dairy farming, with a modest amount of secondary businesses. Medge says: "My wife and my daughter live in Mumbai, my mum lives in the village of Vajori in Rajgurunagar, which is in the Pune district, so she can tend the fields. There's also my uncle, who was my father's brother, and his son. Those who are able to work can also live in Bombay, while those who can't work stay in the village".

Mukadam, one of the NMTBSCT *dabbawalas*, explains his family's geographical situation:

> I live in Goregaon with my family, with the brothers who live here. Two are *dabbawalas*, one is a BST [local Bombay bus service] bus driver. He used to be a *dabbawala*, but now he drives buses. I have two daughters and a son, but my son is seventeen and studies. My other brother and his family stayed in the village because there's nobody else to tend the fields, so he's there with my mother who is alone since my father died.

Work in the fields does not ensure a steady income, hence the desire to migrate to a big city and guarantee the family additional earnings. Most *dabbawala* families do not move to the city as a group. Family members stay in the village, where they continue to farm small plots of land that they own or rent. This migration from the countryside to the city developed with the same dynamics described by cumulative causation theories of contemporary migrations, particularly "chain migration".[48] Sporadic chances to cope with the increased family risks which arise from the effects that metropolitan development triggers on the outskirts will gradually consolidate into a chain migration phenomena facilitated by strengthened family networks and shared geographical origin. The consolidation of migratory movements contributes to the stabilisation of flows because the families of the migrants eventually enjoy an enhanced income and elevation of their social status. Conversely, the families of those who cannot or will not emigrate are the victims of increasing relative poverty.[49]

In the case of the Mumbai *dabbawalas* the existence of a migration chain with a considerable history has actually helped build a culture of migration into urban areas with unique features — for instance, the conservation of an ethic founded in rural virtues combined with the encouragement

---

48   In reference to the analytical prospects summed up in the so-called new economic theory of migration, see Douglas S. Massey, Joaquin Arango, Graeme Hugo, Ali Kouaouci, Adela Pellegrino and J. Edward Taylor, *Worlds in Motion: Understanding International Migration at the End of the Millennium* (Oxford: Oxford University Press, 1998). By "chain migration" I mean the migratory phenomena in which both strong and weak social bonds play a key role in perpetuating a migratory flow from a specific area of origin to one (or more) target areas, ensuring a certain stability to these flows even over many generations.

49   "Relative poverty" or "relative deprivation" is a concept introduced in economics by Oded Stark (see Massey et al, 1998) to explain the many spates of emigration. For Stark, it is above all the tension between low (and slow-growing) incomes and high (fast-growing) incomes that generates and nurtures the migratory tendency.

of new migration from villages around Pune towards the metropolis. One NMTBSCT *dabbawala* describes his employment history and patterns of migration:

> I've been a *dabbawala* for 25 or 30 years. I come from the village and before I was a *dabbawala* I worked in a paper factory there: Auto Plane Paper. I also worked in the fields: I've done everything. But there wasn't enough to live on, so I came here. Sometimes there's rain in the village, sometimes there's none, and sometimes there's too much. That's how farm work is. But I'm not the only one—lots of people from my village have left the fields and gone to do other work. At home now there are one or two men for the fields. The others all do other work. There's no money. What can we do? We go to the city to do other work. I like working in the fields but you can't fill a whole family's stomachs with it. It's a bad business but it's the same for everyone.

Another tells a similar story, in which his family's income is at the mercy of unpredictable farming conditions and increasingly stretched resources:

> I have two sons. One is studying and he's just finished class twelve. He's working in a clothing store at the moment. The other one works in an airline ticket office. When I retire, I'm going back to the village. There's no pension here so you have to work to get the money you need for the future. The association doesn't give you anything. You just work while you can and then you go back to the village. There's no support. Back at the village I've a home and fields. The reason we're here is because we don't get rain on time now. Before the fields gave crops, but not anymore. The new generation has increased. A village used to have two men, now there are two hundred, but the fields haven't increased, they're still the same. Land ownership doesn't increase: the population grows but an acre of land is just an acre of land. A man has two sons, two sons have four sons, then they have to move away to work, otherwise who would leave their villages and their ways! Now, if you're working in an office, they tell you 'You have to dress this way!', but our tiffin work is not like that. Our customs are the same as back in the village. We work in the same way. That's why I like it.

The heavy exploitation of the territory, including excessive deforestation in recent years, has significantly deteriorated the conditions of agricultural work. Drought and torrential rain have eroded arable land, rendering it infertile and difficult to work. All this has contributed to increasing emigration towards Mumbai.

A number of testimonies indicate that some *dabbawalas* identify themselves with the Kunbi *jati*, a caste found in Pune who work the land (Kunbis and Marathas are closely approximated, and in the 1931 census they were classified as one category).[50] There is a possibility that before enlisting in Shivaji's army the ancestors of the *dabbawalas* were Vaishya *varna* and that at the time the term "Maratha" was simply an ethno-linguistic designation identifying Marathi-speaking people. It seems that the intense relationship between Shivaji and the Varkari movement allowed the Kunbis to join its ranks and undertake a *"varna* leap". In this way, they were admitted to the Vaishya *varna*, and this was when the term Maratha began referring to a caste.[51] Thus, the term Maratha identified first the people recruited as warriors and then became an explicit caste indicator, subsequently reproduced and reinforced through intermarriage. Medge explains:

> We *dabbawalas* are farmers from the same rural area and we're all part of Varkari Sampradaya. We have great consideration of God and the *sadhus*. We're illiterate, we don't know how to read and write, so we're not able to do office work. We don't speak English and we have to learn Hindi when we get to Bombay. We have to study it because it's our national language, so we can communicate with customers. In our Sampradaya we consider food to be like a divinity, we wear God's garland. Serving human beings, serving is like rendering service to God, like meeting God. Sampradaya people are all vegetarian and we even take food to young children, to primary and secondary schools, because their parents work in offices and they order food for their children from the *dabbawalas*. "Hindu Kshatriya" — Shivaji is a soldier! To do hard work, you have to be Kshatriya. In our Sampradaya we revere *sadhus* deeply, so that our work goes well. Customer satisfaction, good service, are what a Varkari Sampradaya offers. You need a Kshatriya for hard work.

This brief review of how the Maratha caste developed is just the tip of the iceberg: it is in no way a complete picture of *dabbawala* caste organisation, because some workers have different origins from the Pune area. It is significant, however, that most of the *dabbawalas* come from the same territorial and cultural backgrounds as did most of the Bombay cotton mill workers, which facilitated the development and success of their delivery service. In the same way as other caste associations, the *dabbawalas* have built their business on the basis of

---

50   I am indebted to Ms Gauri Pathak who conducted research into Mumbai *dabbawala* management (Gauri Pathak, personal correspondence). For a complete description of Kunbi habits and customs, refer to Robert Vane Russell, *The Tribes and Castes of the Central Provinces of India*, 4 vols. (London: Macmillan, 1916; repr. New Delhi: Asian Educational Services, 1993).

51   Ibid.

existing linguistic, regional and caste ties, using these common factors as a resource that would generate capital. In Mumbai today, those who are collectively known and recognised as "Marathas" cannot actually be traced precisely to a specific *jati*. The reason is the gradual erosion of the kinship system organised according to the traditional tenets of intermarriage and shared eating habits that underpin the caste concept, following increased rural migration to the city.

Historically, caste identity was reinforced by conflict and antagonism between Brahmins and Kshatriyas in nineteenth- and twentieth-century public Hindu debates. Brahmins were progressively seen as educated but arrogant, and came to be represented as an effeminate, decadent expression of urban culture. In this way they were a total contrast to the basic Maratha Kshatriya *varna* values of warfare, rural virtues (honesty, frugality, humility, decorum, etc), devotion to the Bhakti movement and emphasis on a regional, suburban background.[52] In Mumbai the significance of belonging to a caste increased insofar as it gave access to overlapping networks. These networks were made acceptable and expendable because of shared ideals of rural virtues, economic opportunities, local alliances, and group affinities that emerge on each occasion thanks to common interests and migration processes.[53]

---

52 Hansen (2001), pp. 30–31.

53 Ibid. I would, however, point out that if it is true that the caste system informs and organises Indian society, it is equally true that independent India was established as a democratic, socialist and secular state. The fundamental principles of the Indian constitution guarantee the equality of all citizens before the law (art. 14); prohibition of all forms of discrimination on the basis of religion, caste, race, sex and place of birth (art. 15); the right to freedom of worship (art. 25); the right to receive an education in one's own language, writing and culture (art. 29), the promotion of social mobility for disadvantaged castes (art. 16–17). This policy, with appropriate adjustments, has been applied by all state governments in the Union of India, which enjoy a certain degree of legislative autonomy with respect to the inclusion of some groups in the list of disadvantaged social categories that may take advantage of specific social promotion policies. The Indian government, following the typology of Indian communities that had been developed by the British colonizers, classified the most disadvantaged social categories in three groups. The first was called the "Scheduled Castes" and included all the communities of so-called "untouchables", who now define themselves as "Dalits" ("downtrodden") in modern India; the second, called the "Scheduled Tribes", included Adivasi (aboriginal) communities, i.e. non-Hindu populations living in the forests and mountains of the country's remotest areas; the third groups was called "Other Backward Classes" or simply "Backward Classes" and included many castes that were part of the Shudra, nomads and tribes traditionally known as brigands. See the Constitution of India: http://lawmin.nic.in/olwing/coi/coi-english/coi-indexenglish.htm [accessed 29 June 2012].

Gastrosemantic aspects

Food as a cosmic principle

> *Dabbawalas* consider their work as performing puja. I'm very devout: I have
> a lot of faith in God. Serving food [the speaker used the Sanskrit word anna]
> is a very important thing. The fact is, if you don't bring the food on time, if
> the *dabba* is late for lunchtime, it's a sad moment for the person who doesn't
> get the food they're waiting for. When it's lunchtime, the customer will be
> hungry. So our job is to deliver the *dabba* and if you arrive at lunchtime, the
> customer is satisfied; they get their food and they're happy. That's the work,
> you deliver food to others, so it's good work.
>
> —Director at the NMTBSTC

The NMTBSCT meal delivery service is founded on the unique role
that food plays in Indian culture. The concept of "gastrosemantics" is
used to define this value that, according to Indian anthropologist
Ravindra S. Khare, indicates "a culture's distinct capacity to signify,
experience, systematise, philosophise, and communicate with food and
food practices by pressing appropriate linguistic and cultural devices".[54]
Khare's definition points to the pivotal role played by food in India,
and is useful for delineating the ritual practices, social behaviour and
theological speculations linked to it. Food in India expresses a multitude
of classifications—from satisfying daily biological needs to defining social
and family relationships, economic transactions, hierarchical boundaries,
and ethical and legal systems.[55] Food may be approximated to linguistics,
aesthetics and grammar for its abstract language; on the other hand, it
is a tangible, physical, material substance. This turns it into a cluster of
moral meanings and expressions that reflect the needs of the body and its
aspiration to spiritual liberation.[56]

This section does not claim to give a detailed picture of all the different
meanings that food acquires in India and its many different cultures. Rather
it attempts to define the essence and cultural experience that food evokes
among Indians. Referring back to the work by Khare, the term Hindu is used

---

54   Ravindra S. Khare, "Food with Saints: An Aspect of Hindu Gastrosemantics", in *The
      Eternal Food: Gastronomic Ideas and Experiences of Hindus and Buddhists*, ed. by Ravindra
      S. Khare (Albany: State University of New York Press, 1992), pp. 27–52 (p. 44).

55   Patrick Olivelle, "Food in India", *Journal of Indian Philosophy*, 23 (1995), 367–80.

56   Ravindra S. Khare, "Introduction", in *The Eternal Food: Gastronomic Ideas and Experiences
      of Hindus and Buddhists*, ed. by Ravindra S. Khare (Albany: State University of New York
      Press, 1992), pp. 1–26 (p. 1).

to indicate various traditions that share a common historical civilisation path, so even talking about "Hindu food" becomes infinitely complicated.[57] There is no attempt here to standardise the various cultural, ethnic, religious and linguistic currents present in India, only the desire to find a common denominator. Although Hindus, Buddhists and Jains (to name but a few of the many beliefs found in India) share similar food practices, each group has its own gastrosemantics: food and the act of eating it are a multiple but uniform Hindu "Ultimate Reality"; the Jains are subject to strict austerity and denial; Buddhists follow a principle of moderation.[58]

Although it can be asserted that every community in India has different food ethics, it is still possible to trace a common origin when investigating ancient beliefs and the practices of the various groups speaking Aryan languages. The Aryan peoples settled on the northern Indian plain in about 1500 BC, arriving from central Asia, and their beliefs spread mostly in that part of India.[59] Their food model was based on sheep rearing and agriculture, and was underpinned by a philosophical consideration of eating. Food was not only a form of subsistence but also a fundamental part of the great Aryan cosmic moral circle where those who eat—and the

---

57   One needs to bear in mind Michelguglielmo Torri's point that Indian civilisation should not be identified with the Hindu religion which, in turn, is not defined by discussion of a limited number of texts that thus provide an image of a uniform doctrine. There is a no hard, immutable core in the civilisations because they are all subject to ongoing change through history and that makes it much more complicated to describe them by the use of categories. See Torri (2000), pp. xiv–xv.

58   For a description of the different food styles, see K. T. Achaya, *Indian Food: A Historical Companion* (New Delhi: Oxford University Press, 1994).

59   There are several theories about Aryan migration to India of which the most prevalent describes a violent foray from Central Asia to the north of India, which might also, as Gavin Flood suggests, reflect on the social structure of Europe where this theory was developed. I think it should be stressed that, over and above the different interpretations, the origin of the Aryans has become important for ideological reasons, both in India and in Europe. In the latter, the "invention" of an Aryan race was the basis of Nazi racist ideology; in India, the Hindu nationalist movement recently embraced the autochthonous origin theory to state that the only real inhabitants of India are the descendants of the ancient Aryans, who professed the Hindu religion. Then there are lay groups and the Marxist Left who identify with the migration from Central Asia so they can say that India has been characterised by the presence of many ethnic groups since time immemorial. For an introduction to Aryan migration, see Flood (1996); and Léon Poliakov, *The Aryan Myth: A History of Racist and Nationalistic Ideas in Europe* (New York: Basic Books, 1974); and especially Torri (2000). For the Indian thoughts on the theme, see Romila Thapar, "The Theory of Aryan Race and India: History and Politics", *Social Scientist*, 24 (1996), 3–29. For the relationship between Aryan civilisation and food, see K. T. Achaya, "In India: civiltà pre-ariana and ariana" in *Storia e geografia dell'alimentazione*, ed. by Massimo Montanari and Françoise Sabban, 2 vols. (Turin: Utet, 2006), pp. 144–52.

food they eat—must be in harmony with the Universe. The food ingested, in relation to this harmony, gives rise to three major transformations: faeces, meat and *manas* or thought, which is the most precious transformation.[60] In Sylvain Lévi's opinion, Aryan or Vedic culture was characterised by a strong element of violence generated by the frequent use of meat in individual sustenance. Lévi defines the culture as "brutal" and "materialistic", and its most becoming expression was in the leitmotif of the food and the eaters.[61]

Wendy Doniger has suggested that the nutritional chain describes the order of the species: "what might appear as a culinary metaphor was really meant as a descriptive account of the natural and social world as organised in a hierarchically ordered food chain".[62] In other words, each species eats in proportion to its strength: from big to small, from strong to weak. The linear sequence described delimits a social space, which is reflected in natural space and in ritual sacrifice, whose expression renews the scale of values. Vedic norms were overturned with the spread of the figure of the renouncer in about the sixth century BC, and were conveyed by religious currents within the Hindus, like the Bhakti, who placed emphasis on service and love.

Great spiritual masters like Buddha and Mahavira (the highest authority of Jainism) promoted a purely vegetarian diet in the fifth century BC.[63] Vegetarianism (as well as non-violence), now considered to be the utmost expression of spirituality, was a revolution in Indian society because abstaining from consumption of meat became synonymous with purity and the marker of a true reversal of social values.[64] The new diet

---

60   K. T. Achaya, *A Historical Dictionary of Indian Food* (New Delhi: Oxford University Press, 1998); and Achaya (1994).

61   Sylvain Lévi, *La Doctrine du sacrifice dans les Brahmanas* (Paris: Ernest Leroux, 1898). For complete comments on the subject, see Brian K. Smith, "Eaters, Food and Social Hierarchy in Ancient India: A Dietary Guide to a Revolution of Values", *Journal of the American Academy of Religion*, 58 (1990), 177–206.

62   Wendy Doniger (ed.), *The Laws of Manu* (London: Penguin, 1991).

63   See Achaya (2006).

64   Ahimsa is one of the Jain religion's fundamental principles and has become one of the essential rules of pan-Indian thought. The principle is respected by Jain monks and includes strict rules about food: they cannot eat after sunset, when a living organism could be swallowed in the darkness; they cannot cook food, they only accept it from devotees of the faith, and only leftovers, not food cooked specifically for them. To prepare for asceticism the monks perform different fasts, which may be the progressive limitation of some foods, the exclusion of certain foods (apart from those already normally prohibited: meat, honey, some vegetables, unripe fruit, alcohol), or even a complete rejection of any kind of nourishment to the point of suicide by starvation. See Carlo Della Casa, *Il Gianismo* (Turin: Bollati Boringhieri, 1993).

was not only an expression of a different cuisine, but a new cultural model and worldview.[65] Paradoxically, however, the vegetarian choice promoted by Jains and Buddhists strengthened open discrimination against the Aboriginal peoples who lived by hunting and gathering. Through an act of political opportunism they were relegated to the margins of society, deemed as impure and classified as Untouchables of the Dalit caste.

Sanskrit literature also addresses considerations regarding food and sees the placid, generous cow as the quintessential symbol of restraint in the consumption of meat.[66] Athraveda, for example, emphasises how the cow should be sacrificed only if sterile. The Brahmins and subsequent Upanishads raise doubts as to the use of ritual sacrifices with animal victims.[67] The new spiritual orientation maps out the production and circulation of food applying cosmic logic. The Upanishads affirm that food is a manifestation of Brahma, the Ultimate Reality, and that it influences a Hindu's interior life to the point of controlling development from one birth to another. For this reason there are multiple food classification charts to establish appropriate roles for nutrition practices. It is essential to specify the contexts, conditions, and quality of the food to be consumed or avoided,

---

65 Indian states, with the exception of the states of Bengal, Kerala, Nagaland and Meghalaya, have integrated their laws with the principles of vegetarian ethics. Article 48 of India's Constitution prohibits the slaughter of cows, calves and other milk or draught cattle. Article 48 in its entirety states: "*Organisation of agriculture and animal husbandry*. The State shall endeavour to organise agriculture and animal husbandry on modern and scientific lines and shall, in particular, take steps for preserving and improving the breeds, and prohibiting the slaughter of cows and calves, and other milk and draught cattle". See the Constitution of India, Ministry of Law and Justice, Government of India (2011), available at http://indiacode.nic.in/coiweb/welcome.html [accessed 29 June 2012].

66 In addition to religious explanations, there are also demographic considerations that have given rise to different interpretations of the prohibition of eating cow flesh. These include the demographic growth of Vedic society in 1800–800 BC, leading to a reduction of meat consumption per capita and a trend for a diet consisting of cereals, vegetables, and dairy products. See Massimo Livi Bacci, *Popolazione e alimentazione. Saggio sulla storia demografica europea* (Bologna: Il Mulino, 1993). Also, according to Marvin Harris, cows in India had core tasks in the entire agricultural cycle: they drew the plough to turn the heavy Ganges plain earth, so it was therefore preferable to use them for agricultural work and not as a source of nutrition. See Marvin Harris, *Good to Eat: Riddles of Food and Culture* (Long Grove, IL: Waveland, 1985).

67 Today, Hindus still worship cows and bulls as deities: in Hindu mythology Shiva rides the bull Nandi, and Krishna, protector of children and avatar of Vishnu, is represented as a cattle herder. This has become a key element of Hindu identity, especially in the thinking of the Indian right-wing, to the point of discrediting historical sources and the works of the Sanskrit scholars who documented livestock sacrifice as both a ritual and a source of livelihood.

because the inner state of the being in this world and beyond is intimately connected to what a person eats.[68] This is *dharma*, the cosmic moral order that regulates food availability for all creatures and at the same time is also expressed through complex social distinctions and rituals.[69] This "talking food", to use Khare's phrase,[70] culminates in the production of a non-dichotomised bond between the creator, the body, and the "I", evolving from a need rooted in materiality into an expression of the person's interior life: from generative commodity to cosmic ideal.

This complex universe is seen in the daily lives of people at home and at work, in various ways, since ethical and religious principles are still very much alive in earthly life. In the pragmatic social dimension there is constant interaction between the human and the divine, and it is not limited to the places set aside for worship.[71] Gavin Flood makes an interesting distinction between public and soteriological religion: if the latter addresses the individual and their salvation, the former is "the regulation of communities, the ritual structuring of a person's passage through life and the successful transition, at death, to another world" and "concerned with legitimising hierarchical social relationships and propitiating deities".[72] For the devotee, food is a comprehensive, delicate language marked by a broad spectrum of cosmic, social, karmic, spiritual and emotional messages: a language that speaks both through the choice of ingredients and beyond the boundaries of materiality.

Food-related holy practices

Eaten or even just handled, food is known to have a dual action. First, it provides biological support for the body's vital processes; secondly, it acts to achieve experiences of "a more subtle nature" because it allows for the amplification of spiritual perceptions by linking belly, mind and soul.[73] Foods, especially those considered "good" or "pure", allow humans to renew primeval harmony among all nature's creatures. The act of eating

---

68   Khare (1996).
69   Doniger (1991).
70   Khare (1996), p. 18.
71   Pier Giorgio Solinas, "Soggetti estesi e relazioni multiple. Questioni di antropologia indianista", *Società Degli Individui*, 25 (2006), available at http://www.antropologica. unisi.it/images/a/ad/L'in-dividuo.pdf [accessed 29 June 2012]. DOI: 10.1400/65102
72   Flood (1996), pp. 13–16. See also Richard Gombrich, *Theravada Buddhism: A Social History from Ancient Benares to Modern Colombo* (New York: Routledge, 1988).
73   Mario Bacchiega, *Il pasto sacro* (Padova: Cidema, 1971), p. 137.

actually implies the destruction of a being and awareness of this violence, directly linked to death. This means that the act in itself is problematic, especially with regard to the consumption of meat, which requires the eater to perform a purification ritual to expiate the guilt linked to violence and restore the bond that has been broken by the killing. In turn, the ritual legitimises the eternal nutrition cycle sequence.

Moreover, Nature—perceived as a deity—suffers predatory human actions: agriculture and the gathering of wild berries are part of an exploitation system that causes humanity to seek a justification for its acts. This occurs through rituals celebrating the perpetuation of life by a sacrifice that appears in different forms and may involve people, animals, nature and divinities. The sacrifice generates a pact between living beings, linking them in an ongoing sequence of life and death. The banquet, which everyone attends although in different ways (there are those who eat and those who are eaten), becomes the privileged channel for achieving an inner opening towards the hallowed and greater communion with the divine. Notions underlying the classification of foods thus assume primary importance, since the level of purity brings spiritual elevation of different intensities. This classification differs depending on the various cultures that developed in India.

The spiritual rules connected to nutrition are extensively articulated in the practice of Ayurveda, a term of Sanskrit origin meaning knowledge of life, composed of the word *ayus* (life) and *veda* (knowledge).[74] Ayurvedic medicine does not separate the health of the body from that of the mind and aims to restore the balance among all the components that reflect an individual's health: care of organs, psyche and soul; and even the relationship the person has with the environment, relations with the family and with the world at large. For a healthy person, the purpose of Ayurvedic science is to pursue and achieve four objectives in their existence:

- *dharma* (achieving wellness through decorous living, including respect for justice and morality);
- *artha* (achieving a good standard of living, while respecting the rules of *dharma*);

---

74 The main sources of Ayurveda are the *Caraka Samhita* (*Caraka Collection*) describing the legend of the origins of Ayurveda and the *Susruta Samhita* (*Susruta Collection*), which is especially important for surgery. See Isabella Miavaldi, *La cucina ayurvedica* (Milan: Xenia, 1999).

- *kama* (the satisfaction of worldly desires, passion and love); and
- *moksha* (attainment of salvation and liberation from the cycle of birth and death acquired through consciousness of the existence of God).[75]

These objectives allow equilibrium to be found both for the inner self and with the environment, or better still, a balance between the macrocosm and the microcosm (the body), the latter reflected in the former. Disease is thought to be the result of a breach of this balance and Ayurveda, as a therapeutic science, recommends proper nutrition and yoga as supports to maintaining good health through inner peace, transcending the senses and releasing the bonds with matter.[76]

Human well-being relies on food and its digestion because the body grows and develops depending on how it is fed. Ayurveda places ample attention on food quality, properties of raw materials, and the processing triggered by *agni*, the body's digestive fire. Foods, like everything existing in nature, have three qualities or *gunas*: *sattvic* (pure), *rajasic* (overexcited) and *tamasic* (rotten). *Sattvic* or pure foods are the best for eating correctly and they include fresh vegetables, fresh fruit, cereals, natural sweeteners, mother's milk, butter, ghee, cold pressed oil and shoots. The *Bhagavad Gita* states that "those foods that enhance life, purity, strength, health, joy and happiness, which are tasty and oily, nutritious and pleasing, are dear to *sattvic* people" (XVII, 8). Only moderate use should be made of *rajasic* foods, which often have a spicy taste and include fermented foods, garlic, cheese, sugar, salt, coffee, chocolate, and very strong spices and herbs. The *Bhagavad Gita* also states that "bitter, acidic, salty, overly hot, pungent, dry or spicy foods are preferred by *rajasic* persons and generate pain, distress and sickness" (XVII, 9). Finally there is *tamasic*: damaged or improperly cooked foods, such as fried and frozen foods, and those treated with preservatives or microwaved. *Tamasic* foods also include mushrooms, meat, fish, onions, garlic and substances like alcohol and tobacco, all of which should be avoided.[77] The *Bhagavad Gita* reads: "the *tamasic* will prefer food that is tasteless and rotten, which is unclean waste" (XVII, 10).

---

75   Ibid., pp. 12–13.
76   There is an extensive bibliography on the birth and evolution of yoga. For instance: Mircea Eliade, *Yoga, Immortality and Freedom* (London: Routledge, 1958); by the same author *Techniques du Yoga* (Paris: Gallimard, 1948).
77   Miavaldi (1999), p. 38; and Amadea Morningstar and Urmila Desai, *The Ayurvedic Cookbook* (New Delhi: Motilal Banarsidass, 1994). See also Dominik Wujastyk (ed.), *The Roots of Ayurveda: Selections from Sanskrit Medical Writings* (London: Penguin, 2003).

It is also important to appraise the taste of foods, namely their *rasa*, because different flavour nuances have different effects on the human body and, not least of all, on the mood of those who consume them.[78] Ayurveda recommends taking care of the body by eating regularly and taking daily exercise. It is important to wake up early, to thank the Lord, empty the bowels so as to start a new day without the debris of the day before. Equally it is necessary to clean all parts of the body properly, with particular attention to the "nine gates". The body is seen as a temple with nine entrances to the outside world: eyes (*netra*), nose (*neti*), ears (*karna*), mouth (*mansuya*), vagina (*yoni*), penis (*lingam*), anus (*mula*), navel (*nabi*) and the top of the head (*brahmarandra*). Massage and meditation are integral to this care as these practices help ward off negative moods so as to avoid anger, envy, greed, jealousy and harming oneself and others.[79]

What seem to be just rules for proper nutrition and lifestyle actually have a close bearing on the sphere of the sacred. Food is *anna*, the first Sanskrit word to designate a Brahmin. Everything in the universe is food and interior growth depends on the ability to eat and digest the food that is our lives.[80] In particular, the choice of foods to combine tends to avoid the juxtaposition of principles that lack equilibrium and so would lead to inner disharmony if consumed.[81] Those who want to keep good *karma* will avoid combining animal food (obtained by the act of violence intrinsic to slaughter) with vegetables (a spontaneous gift of nature).[82] The rule of food harmony tries to achieve an inner balance intended to come close to the harmony of the universe. So, the purer the food is, the more the body acquires its characteristics, enhancing spiritual elevation. This is because foods are not a simple matter: they are a vehicle of subtle information, energies, manifestations of the primordial vibrating energy called *Prana*.[83] Thus, when people eat, they take the Prana contained in the food and

---

78   *Rasa* literally means "lymph" or "edible juice".
79   Gabriella Cella Al-Chamali, *Ayurveda e salute. Come curarsi con l'antica medicina indiana* (Milan: Sonzogno, 1994).
80   David Frawley, *Yoga and Ayurveda: Self-Healing and Self-Realisation* (Delhi: Motilal Banarsidass, 2000).
81   Bacchiega (1971), p. 143.
82   This hypothesis seems universal within food rules imposed by different religions. In Judaism, Deuteronomy (Chapter XIV), there is the famous rule that prohibits cooking the kid in its mother's milk, so forbids combining the violence of slaughtered meat with the mild, pure and gentle gift or milk.
83   Deborah Pavanello, *Cibo per l'anima. Il significato delle prescrizioni alimentari nella grandi religioni* (Rome: Mediterranee, 2006).

circulate it around the body through seven energy centres called *chakras* (specifically: *muladhara, swadhisthana, manipura, anahata, visshuddhi, ajna* and *sahasrara*), which govern various functions and organs. When *Prana* enters a *chakra* through food, organs come into contact with it and it takes a specific name, becoming *apana* at the level of the first *chakra*, controlling excreta; at the second chakra it is *viana*, which regulates blood circulation; as *samana* at the third chakra it regulates the digestive process; *prana* (without a capital) at the level of the heart chakra, controls the respiratory process; lastly there is the fifth chakra, *udana*, which controls the diaphragm.[84]

Action (or non-action) that arises from the fermentation of pure foods within the human belly is thought to bring vital thought processes closer to the cosmos and to the divine. In this way, foods become mediators that can absorb and convey the subtle energies that connect human beings. As mediators they permit the transfer and the "initiatory succession" of energies that offer the eater the possibility of being grafted, the Vedas say, into "cosmic light". Consequently, the meal is a rite, a moment of exploration, of learning, but also of intense rapport with the other, the Absolute, an act that allows their "realisation in existing and in strength".[85] But there is no dichotomy between what is eaten and the eater because, as Mario Bacchiega says, "everything has been eaten" and "the eater and the eaten are the same thing [...] really there is neither eaten nor eater".[86] The extraordinary mystical chant: "I am food, I am food, I who am food, I eat the eater of the food!" expresses the overcoming of the tension between knowledge and love in the symbol of food, because here human and divine action need each other. In this cosmic metabolism, profound unity is achieved: eat the other and be food for the other.[87]

Food is not only limited to the sphere of the sacred, however: it has an aesthetic, popular aspect that reflects daily life. Indians generally do not possess an in-depth knowledge of all expressions and characteristics that food plays in the culture and spirituality of India, but they internalise the guidelines of nutrition (including the beliefs, functions and traditions related to it) from the families in which they are raised. Often this is represented by the daily rapport that women have with the handling of food and with children. So even

---

84   Ibid., pp. 127–28.
85   Bacchiega (1971), p. 270.
86   Ibid., p. 271.
87   Raimundo Panikkar, *The Vedic Experience: Mantramanjari* (New Delhi: Sanctum, 1977), pp. 306–07.

if *haute cuisine* in traditional patriarchal society is the expression of a public and ritual act and the domain of male Brahmins, it is usually the woman who is the leading player in cooking and preparing meals.[88]

Food in women's everyday lives

Any discussion involving food must take into account the role of women in the preparation of meals and the close bond between women and the act of cooking. Indian women today enjoy a wide scope of action, but traditionally their place has always been in the home, particularly in the kitchen, where the family shrine was, and still is, placed. The kitchen is the heart of the Hindu home, which is kept as far as possible from sources of contamination such as sleeping quarters or the room where visitors are received. Before entering the kitchen, the cook must clean herself of any contamination that may come from the outside and change clothes, because the purpose of the food preparation process is not only to produce foods that keep the body alive, but to merge the cultural properties of the food transformed by the cooking process itself with those of the people who eat it.[89]

The woman plays a vital role in cooking for the family and providing food for the gods. This act requires knowledge of the preferences of the deities and those preferences, with relevant recipes, are handed down from mother to daughter. Usually, however, the offering to the gods, called *prasad*, includes rice boiled in milk, small cakes, a stick of incense and a garland of flowers. After the god has been fed, the leftovers are redistributed among family members. As mentioned previously, the cow, and cattle in general, have a crucial importance in the Hindu religion and in Indian culture. Women are often privileged custodians who wash cows and decorate them for religious ceremonies and festivals. Fresh cow's milk is used to wash the statues in family temples, while dry dung is used as fuel for cooking (especially in the villages) and to clean the floors of the house and the kitchen. The kitchen floor is used as a table to be set before sitting down for meals. It is always the women who bring the meal to their families and while serving they "give". Even in modern times, women–especially the

---

88 See Kirti Narayan Chaudhuri, *Asia before Europe: Economy and Civilisation of the Indian Ocean from the Rise of Islam to 1750* (Cambridge: Cambridge University Press, 1991); and Charles Malamoud, *Cooking the World: Ritual and Thought in Ancient India* (Delhi: Oxford University Press, 1996).

89 See Achaya (1994).

older generation—in the villages do not eat with the family, but only after everyone else has finished.

Cooking is not limited to the scope of the food but expresses a wider range of meanings: a woman's inner heat (a force known as *shakti*) is said to be ten times higher than that of a man. David Smith writes "it is this force that enables them to give birth, to as it were cook the foetus in the same way as they cook the food that maintains the life of the family they have given birth to. The husband is born again in the son that originates from his wife's womb. Husband and children are all given life and physical sustenance by the wife".[90] The birth of a child, in particular a son, confirms the woman in her role within the family and through generational continuity she saves her husband from the condition of being a man without descendants. The son will ensure nourishment to his father's spirit after his death. The woman also renews the act of cooking in every act of sexual intercourse because through her heat she "cooks" the male member.

This symbolism, related to women's bodies, is not immune to the pure/impure distinction that regulates certain times.[91] During the menstrual cycle, for example, women—especially of the higher castes—do not cook or enter the temple.[92] The mother's preference for pure foods ensures her offspring remain healthy in line with the dietary requirements laid down by *dharma*. Those individuals who are fed pure foods (mainly vegetables) are reborn in a high social status; conversely, those who have eaten animal flesh or pursued less discriminating eating habits, will be reborn in a lower *varna*.[93]

This is in no way an exhaustive description of the relationship between the female body and cooking in the broadest sense, but it is useful for understanding the context in which women once operated, and still do, although they now enjoy a freedom that releases them from some traditional domestic constraints, especially if they live in large cities and

---

90   David Smith, *Hinduism and Modernity* (Oxford: Blackwell, 2003), p. 110.

91   A clarification is required here of the terms "pure" and "impure". These are not necessarily used as moral judgements: Indians believe that they refer to a physical condition. Impurity is represented by nature and its manifestations, which may lead to contamination and must be limited in order to prevent infection or disease. It is clear that this dichotomy also leads to social hierarchical distinctions that define the relationships between the castes, whereby upper castes maintain their purity by avoiding contact with impurities, which are passed on to lower castes.

92   See Smith (2003).

93   Mario Piantelli, "Lo hinduismo. I. Testi e dottrine", in *Hinduismo*, ed. by Giovanni Filoramo (Rome: Laterza, 2002), p. 90.

are relatively wealthy. These constraints do, however, fall upon the work of women of more humble extraction, who perform domestic services in middle-class houses (from cleaning to managing the kitchen). Sociologist Barbara Ehrenreich and anthropologist Arlie Russell Hochschild have shown how, in the battle for equality and the right to self-determination, feminism often hides a form of exploitation carried out by richer, perhaps career-oriented women, towards less-educated or poorer women. Rather than men taking on traditional female tasks as more women head to the workplace, domestic duties are offloaded onto other women.[94]

The caste hierarchy of food

Food transactions are not only relevant in gender differences; McKim Marriott believes that they could be a primary device for explaining caste organisation.[95] The food reflects caste differences through a series of rules with which diners must comply scrupulously to avoid contacts that could render both the food and people who are eating impure. Written rules list the different types of impurities (for example, the moment of birth and death, or the performance of certain manual tasks that involve contact with unclean elements), but generally the act of eating is a far more vulnerable time than others and should be approached with great care. Food rules that affect ordinary meal consumption presume an attention to ingredients (vegetable foods are preferable to animal products); to cooking (preferring food fried in ghee or clarified butter to boiling); place (the eating place should be as far away as possible from possible sources of contamination); cookware (preferring the use of copper or aluminium pans to terracotta because they can be washed with greater ease and do not accumulate residues in porous cavities). Every gesture during the meal must be controlled to avoid making the food inedible. For example, vicinity to lower caste individuals should be avoided, as should the presence of animals or contact with human saliva. Despite being produced by the body, saliva is regarded as "alien" to it because it is a vehicle of potentially harmful

---

94   Barbara Ehrenreich and Arlie Russell Hochschild, *Global Woman: Nannies, Maids, and Sex Workers in the New Economy* (London: Granta, 2003). See also Henrike Donner, *Domestic Goddesses: Maternity, Globalization and Middle-class Identity in Contemporary India* (London: Ashgate, 2008).

95   McKim Marriott, "Caste Ranking and Food Transactions: A Matrix Analysis", in *Structure and Change in Indian Society*, ed. by Milton Singer and Bernard S. Cohn (Chicago: Aldine, 1968), pp. 133–72.

substances, even though it is also synonymous with deep acceptance and belonging to the family group.

By tradition, and especially in the first phase of the marriage, a wife eats her husband's and in-laws' leftovers so as to be integrated into the family. A parent may consume their child's leftovers. Sharing a meal is a family action, because the family eats "from the same hearth".[96] If a member breaks a caste rule (for example, by offending a family member), they will not be accepted at family lunches and reintegration is symbolised by a party in which the offended person offers food to the offending person. This gesture allows all other members to reintegrate the offender into the eating circle.[97]

The rules are stricter for the high castes, in particular for the Brahmins, and impose a series of precautions that do not leave much room for individual freedom.[98] For Brahmins in particular, a vegetarian diet is strictly necessary to maintain caste purity and foods should always be placed safely in sealed containers to avoid being contaminated by an outsider's impure hands. This code of conduct was already noted in the eleventh century by the Arab intellectual Al-Biruni. He describes relations between Hindus and Muslims:

> all their fanaticism is directed against those who do not belong to them — against all foreigners. They call them *mleccha*, i.e. impure, and forbid having any connection with them, by intermarriage or by any other kind of relationship, or by sitting, eating and drinking with them, because thereby, they think they would be polluted. They consider as impure anything that touches the fire and the water of a foreigner; and no household can exist without these two elements. Besides, they never desire that a thing which once has been polluted should be purified and thus recovered [...].[99]

The excerpt highlights how the concept of purity was an element of identification by foreign observers. Food preparation and habits were signals that revealed caste hierarchy and Hindu culture more broadly.

A 1970s study led by Stanley Freed in Shanti Nagar, a village in northern India, showed how the caste hierarchy was based on an asymmetric exchange of food and water. Caste classification is apparent

---

96   Lawrence A. Babb, "The Food of the Gods in Chhattisgarh: Some Structural Features of Hindu Ritual", *Southwestern Journal of Anthropology*, 26 (1970), 287–304 (p. 297).

97   Ibid.

98   Dumont (1996).

99   Al-Biruni, *Alberuni's India: An Account of the Religion, Philosophy, Literature, Geography, Chronology, Astronomy, Customs, Laws, and Astrology of India about A.D. 1030*, trans. by E.C. Sachau (New Delhi: Munshiram Manoharlal, 1992), vol. I, p. 25. My quote is from Smith (2003), p. 73.

in understanding from whom food can and cannot be accepted. The giver is in a hierarchically superior position to the receiver. The asymmetry of the exchange makes use of cooking techniques. For example a *pakka* food, cooked in clarified butter, can be taken by members of higher castes; conversely a *kaccha* food cannot.[100] *Kaccha* and *pakka* mean literally "raw" and "cooked" but the derived sense has a very extensive application and indicates, on one hand, insecurity and imperfection, and on the other soundness and perfection (a notion that contains a hierarchical nuance). The distinction between these two stages is achieved by the use of ghee. In Brahmanic India, sacrificial food is counterpoised to non-sacrificial food, and the former is always cooked, *pakka*, protected and sanctified by the ghee used for frying it. *Pakka* food is less exposed to contamination while *kaccha* food is more fragile and corruptible.

Modern meaning intends both as cooked food but using different techniques: *Kaccha* foods are cooked in water while *pakka* food is essentially fried in ghee and prepared to be consumed outside the home. The cooking sequence is critical in designating one or other type of food. The same principle can be found for water, which retains its purity when kept in a brass jug and can be accepted by members of higher castes when offered by members of lower castes; when it comes from earthenware jugs this is not possible.[101] Food thus becomes the expression of a refined taxonomy, functional in a classification of the universe that reflects natural elements and social orders consistent with the construction of a collective sense. Chaos is opposed by a system of linkages between nature and society, abstractions or cultures, in which humans and worldly objects belong to each other according to a logic of obedience to certain criteria.[102]

Food and the metropolis

In recent years there have been significant transformations in food culture. In the big cities and in diasporic contexts, the caste system softens into

---

100 Malamoud (1996); Achaya (1994).
101 Stanley A. Freed, "Caste Ranking and the Exchange of Food and Water in a North Indian Village", *Anthropological Quarterly*, 43 (1970), 1–13.
102 Brian K. Smith, *Classifying the Universe: The Ancient India Varna System and the Origins of Caste* (New York: Oxford University Press, 1949).

a more fluid stratification where surviving inequalities and differences are based on property, power and prestige. Food also reflects these transformations and, indeed, the current scenario of a metropolis like Mumbai is characterised by a public space dominated by an impressive variety of food, including restaurants that serve any type of food and street vendors offering cheap lunches from their stalls, to eat while standing.[103] The food sold reflects Mumbai's cultural habits and traditions, for instance distinguishing between "hot" and "cold" food, depending on the season it is offered. The distinction does not refer to the temperature of the food or its spiciness, but to the effects the foods have on the body. During the main festivities the dishes offered by street vendors are appropriate to the circumstances and are a good alternative to eating lunch out. Vendors often use ghee instead of water to make food *pakka* (safer). According to Mina Thakur's research on street food in the city of Guwahati, many customers seek Brahmin vendors to be sure they are buying unpolluted food, or they look for vendors of their own caste offering traditional dishes.[104]

In Mumbai, the debate about the regulation of street vendors can get very heated. A recent investigation by Jonathan Shapiro Anjaria noted complaints from city associations about vendors selling street food. These vendors are seen as the symbol of a metropolitan space that escapes the control of the authorities and of the flow of migrants arriving in Mumbai. Often the protests are based on two main elements—language and religion—that reflect the offer of non-vegetarian foods from non-Hindu sellers. In deciding who can sell food on Mumbai territory, the city pursues a nativist policy and distinguishes foods that may or may not be cooked on public land.[105] Often the younger generation call this "junk food", using a common English expression that denotes readily available, cheap food of poor quality. It refers to what students eat for lunch at street stalls, like *bara pav* (a type

---

103   Frank F. Conlon, "Dining Out in Bombay/Mumbai: An Exploration of an Indian City's Public Culture", in *Urban Studies*, ed. by Sujata Patel and Kushal Deb (New Delhi: Oxford University Press, 2006), pp. 390–413; and Rashmi Uday Singh, *Times Food Guide Mumbai 2007* (New Delhi: The Times of India, 2007).

104   The notion of "safe" and "unpolluted" food is obviously to be taken to mean parameters other than those of hygiene and cleanliness. I refer to the rules aimed at preserving caste purity which see the Brahmin caste at the top of the social hierarchy. This position allows them to cook food for all other castes. Irene Tinker, *Street Foods: Urban Food and Employment in Developing Countries* (New York: Oxford University Press, 1997), p. 179.

105   Jonathan Shapiro Anjaria, "Street Hawkers and Public Space in Mumbai", *Economic and Political Weekly*, 27 May 2006, pp. 2140–46.

of bun, cut in half and stuffed with a red or green chilli sauce, potatoes and spices), or *pav bhaji* (a sandwich accompanied by a mixture of vegetables).

The diversified food on offer reflects the demands of a city where new social classes are stratifying and seeking a variety of choices that were unknown a few years ago. Mumbai has undergone a transformation in recent years that entails the growing presence of middle-class people used to eating out in the evening. It is a relatively recent phenomenon, for previously only people with a relatively low income were in the habit of eating outside their homes. The indoor restaurant, different from street food vendors, reflects and promotes a series of changes in public and private contemporary Indian life induced by increased wages and the entry of middle-class women into the working world. These changes were followed by new experiences of conviviality and socialisation, with the consequent modification to spaces, places, and relationships inside the kitchen.

While there are increasing numbers of trendy new restaurants around the city (with many Italian newcomers), multicultural food is not a recent phenomenon in India. There has been a continuous rotation of renewed migratory waves that bring disparate influences to the city's flavours: there is a mix of large colonial Bombay hotels used by the British; the already mentioned *khanawals*; Irani cafés, an important Bombay institution now facing extinction; Western or European-inspired restaurants; and *dhabas* (eateries offering traditional food). These venues offer cuisine typical of the various regions of northern India: Kashmiri; Punjabi; Pahari; Marwari; Rajastani; Lucknawi.

Although regional differences are countless, a typical lunch includes *chappati, paratha* or *poori* (unleavened bread), rice, meat, creamy *dhal* (lentils), meat cooked in ghee or a typical tandoor, kebabs, seasonal vegetables with yogurt, *paneer* (cottage cheese cooked with green vegetables or onions), fresh chillies and fresh tomatoes. Desserts are made of milk, *paneer*, legume flour and white wheat flour cooked with dry walnuts. Lunch is accompanied by drinks like *nimbu pani* (lemonade) and *lassi* (iced yoghurt, sometimes flavoured). Also available are southern Indian dishes from Karnataka, Andhra, Hyderabad, Tamil, Chettinad and Kerala. These cuisines are rice-based (long, short, or round grains) or are simply eaten with lentils cooked to make *uttapam, idli* and *dosa*. Sauces typical of southern India are made from tamarind, coconut, peanuts and *dhal*. The food is served on a *vazhaillai*, a freshly-cut banana leaf. A curious fact related to this "tray" is that the shape of each leaf can be used to identify the food of different communities, a sort of ID card; the arrangement of the food on the leaf also indicates the place of origin.

Mumbaites eat many dishes from nearby Gujarat, a mostly vegetarian diet, which was also widely influenced by Chinese cuisine because of the intense trade that the main port of Surat has always enjoyed with China. There are also typical Maharashtra dishes with their extensive use of fish, and Goan cuisine with its Portuguese influences. Today, however, these are only general categories of catering establishments and the choice of food seems endless. A short stroll in any part of the city will reveal how public space is dominated by the presence of food vendors and places to stop even just for a drink. For example, the middle classes, young students, or professionals from the film world frequent the Barista chain, which offers snacks and coffee in a comfortable setting for chatting or talking business. There are also restaurants serving multi-menus and offering different traditional dishes, or fusions of various food styles known as world cuisine, or kosher Indian, Parsi and "continental" food. Big shopping centres, mainly on the outskirts of the city, have popular food courts offering different types of fare: Punjabi and Udupi; vegetarian and non-vegetarian; Italian ice cream parlours; pizzas; or Chinese and Thai food.

Despite this proliferation of offerings, the constant growth of the *dabbawala* association nonetheless shows that the cosmopolitan side of the city has not homogenised the sensory category of taste. If anything, it has been enhanced seamlessly with the history of the city, which is able to absorb foreign elements by transforming them through a process of domestication. In this way, Mumbai reflects a multifaceted flexibility and through constant negotiation there is not only a commercial ferment that expresses its nature, but also a broader principle of hope for the multitude of people who live there. This multidimensional coexistence is described by Medge:

> Indian people who have the qualifications work in banks, post offices, the railways, aeronautics, are college professors or teachers. Geographically, Bombay is similar to a straight line 72–75 kilometres long. The cost of living in Bombay is high and you cannot live close to offices and places of business. So you go to Bombay to work and live far away. So trains make it easy to move around, local trains. Because of the traffic we use the train and you can get from Virar to Churchgate in an hour. Travelling by train gets you there quicker. There's a local train every three minutes — it may be a fast or a slow train, but you can't get on without trains. Today, Mumbai is home to people

of all castes—Marathi, Muslim, Parsee, Christian, Punjabi, Gujarati, Marvari, foreigners, people coming from South and North India. They all have very different cuisines. If you work in an office at Nariman Point, then you want to choose your own food. Italians eat one way, Indians in another, just like Konkanis, Malvanis, Kuneris, Kolapuris, Solapuris, vegetarians, non-vegetarians, Gujaratis, Punjabis [...] everybody chooses different dishes. Each restaurant can prepare only one type of food. There are so many types of restaurants. Then everyone is accustomed to home food. When you're at home there's a mother, a wife, a daughter, a sister, to cook with their own hands, everyone is accustomed to that food and they don't like restaurant food. If people eat the same food, they lose their tempers. If you work for ten or twenty years, you're the person who brings home the wages, the family prepares the food you like with different spices, without oil or with plenty of oil, whatever way you like, and you'll be served. Because eating well, keeps your mind alert and then it will work just fine, you'll earn well. You will live longer. In Bombay there's so much pollution that everyone's used to home-cooked food but the trains are so busy you can't bring that food with you. That's when you order tiffin.

The *dabbawalas* and the food delivery system

Today the debate on food in India is split around demand (in a country where access to food resources is still deeply marked by social inequality); the right to food for all; and—conversely—the symbolic value associated with caste membership.[106] In relation to demand, one of India's most representative voices, the scientist and intellectual Vandana Shiva, suggests that a plan is under way to use food as a weapon against India and, in general, against developing countries, by subtracting resources like biodiversity and water to local communities. This happens, according to Shiva, through the imposition of rules and laws like the agreement on TRIPS (Trade-Related Intellectual Property Rights) and the General Agreement on Trade, both reached within the World Trade Organisation. The liberalisation of seeds imposed by the World Bank has, for example, allowed multinational companies like Monsanto and Cargill to penetrate the Indian market, buying up the sector's biggest enterprises and persuading farmers to buy very costly fertilisers and hybridised seeds. Nevertheless, this did

---

106 For a full discussion of the concept of "food for all", see Jean Drèze and Amartya Sen, *Hunger and Public Action* (New Delhi: Oxford University Press, 1989). The debate revolves around several key concepts: food sovereignty; food safety; and the right to food. In-depth references can be found on the websites of the FAO or of several NGOs involved in international cooperation projects.

not give the desired results, particularly as far as cotton was concerned, and led to suicide and fatal starvation for many peasants in debt due to the high cost of seed. Shiva believes that low-productivity monocultures and monopolies symbolise the "masculinisation" of agriculture and, together with the patriarchal globalisation project, lead to an increase in violence against women and minority groups. It is a policy of exclusion that bases its legitimacy on the ownership of living beings and plants, conflicting with the shared knowledge promoted by an agriculture based on diversity, decentralisation and ecological methods.[107]

In relation to the symbolic value of food, a recent study undertaken of Udupi vegetarian restaurants, typical of southern India but also widely present in Mumbai, shows how caste-system food prohibitions were present in the city until a few years ago, and it is presumed that the situation still largely continues. Until the 1940s, some "Udupi hotels" had three different entrances: one for orthodox Brahmins, one for liberal Brahmins, and one for non-Brahmins. Today this specification is prohibited by law but can still be found in the owner's surname.[108]

An interview with a cook who works in the kitchens preparing lunches delivered by *dabbawalas* revealed that customer demands are not just about ingredients in the dishes, but also about the rules laid down by the ancient Ayurveda science and spiritual concepts linked to food. The cook is Punjabi but has lived in Mumbai for many years, and after a lifetime spent with her husband (the owner of a small dyeing plant), she decided to work preparing meals. She talked about the attention to healthy eating and home-style cooking (home food) that complies with customer requests for food without, for example, garlic, too many spices or oil. There are no special requirements dictated by caste, but through the choice of ingredients it is possible to reconstruct not only taste trends, but also regional origins and spiritual beliefs of the customer.[109] Despite the diversity of culinary techniques and

---

107   Vandana Shiva, *India Divided: Diversity and Democracy Under Attack* (New York: Seven Stories Press, 2005). By the same author: *Water Wars: Privatization, Pollution, and Profit* (Cambridge, MA: South End Press, 2004); *Earth Democracy: Justice, Sustainability, and Peace* (Cambridge, MA: South End Press, 2005).

108   See Vegard Iversen and P. S. Raghavendra, "What the Signboard Hides: Food, Caste and Employability in Small South Indian Eating Places", *Contributions to Indian Sociology*, 40 (2006), 311–41 DOI: 10.1177/006996670604000302; and U. B. Rajalakshmi, *Udupi Cuisine* (Bangalore: Prism, 2000).

109   For the sake of completeness, I include part of the interview: "My customers ask for food without garlic, that's not too spicy and oily. They're health-conscious people. If anyone has special problems, like they suffer from diabetes, I'm careful with the ingredients

connected recipes, the cook outlined a number of elements that are found all over the country:

- Aroma: the cook must pay attention to the specific characteristics and fragrances of the ingredients. The senses should all be involved in eating. It is thought that aroma allows digestion to begin before the food is consumed.

- Colour: Indian cuisine requires presentation to be a joy for the eyes, with golden lentils, white rice, green vegetables, red condiment, orange meat and spices.

- Flavour: Indian cuisine follows the philosophy of using six flavours in the combination of a complete dish: sweet, bitter, salty, acidic, astringent and sour. Taste is thought to be crucial in preventing disease and in regulating bodily functions.

- Diversity: there are many different recipe versions. Until a few years ago there were no written cookbooks and the majority of food culture was passed down orally. Every housewife then modified the recipes she had learned to her liking and depending on the availability of ingredients, which brought continuous change.

- Structure: the composition of dishes must bring out each food's own characteristics, harmonising the combination of various recipes in a fine balance. The structure is found in *thali*, a typical meal served on a large flat silver or aluminium plate, which includes rice, vegetables, *dhal*, sauces, yogurt, meat or fish (for non-vegetarians), *roti* (a kind of flatbread) and desserts.

From an anthropological standpoint, the main concern is not with caste system-related prohibitions as such, but with what they stand for. As Mary Douglas has suggested, the idea of pollution serves both an instrumental and an expressive purpose. Its instrumental purpose usually translates into an attempt to limit individual behaviour through a set of rules. Yet when their expressive purpose is considered, the same proscriptions appear to be "used as analogies for expressing a general view of the social order".[110] As food

---

even though it's difficult to cook separately just for one person. If someone doesn't want potatoes, I cook a mix of vegetables, but my food is very simple, home food. As the customer's mother or wife were cooking. I'm originally from the Punjab: my mother and aunts taught me to cook. I do those same dishes for my customers".

110 Mary Douglas, *Purity and Danger: An Analysis of Concepts of Pollution and Taboo* (London: Routledge, 1970), p. 4.

taboos emphasise cultural differences and reinforce inter-group boundaries, they impose a specific social order. Such norms, derived from clearly defined notions of what is pure and what is impure, will only have significance within the systemic framework they help to enforce.[111] While specific ideas of pure and impure take hold as codified norms to prevent contamination, they also give shape to the cognitive system people are taught to rely on in order to recognise and differentiate each other's social identity in a shared environment.

Therefore, the symbolic boundaries enforced by food rules do not simply instrumentally distinguish what is edible from what is not, but they also expressively depict a whole hierarchy of social relations, marking out the range of structured kinship bonds, separating ethnic groups and highlighting religious affiliations. The resulting mix of cuisines and ingredients thus maps out a complex "food system" of different eating patterns that mirrors a composite community's daily reproduction of its own social order. Needless to say, in Mumbai this is a very dynamic system that does allow variations, albeit within a recognised, well-entrenched organisational structure.[112]

Within the context of this food system's deeper, expressive symbolism, the Mumbai *dabbawalas'* food delivery organisation can be viewed as a vehicle of purity within what a Marathi poet refers to as the city's "putrid culture", a culture of unavoidable confusion and promiscuity, and a challenge to any conceit of purity preservation.[113] Yet in a city as diverse, shape-shifting, at once life-giving and soul-tainting as Mumbai, food delivery services make it possible to preserve bonds of affection and family roles as one's spouse, parents or relatives prepare their out-of-home family member's meals *at home*. Moreover, the *dabbawala* system enables individuals and their families to uphold nutritional choices that, in terms both of ingredients and taste, adhere to the requirements and norms they most cherish, whether caste-related, status-driven, or in other ways dictated by social, religious or cultural custom. As an added bonus, the system allows its users to avert the risk of ingesting polluted and poor quality food while away from home.

---

111 Ibid., p. 35.
112 The theoretical reference is described in Mary Douglas, *Implicit Meanings* (London: Routledge, 1975).
113 The Marathi poem *Mumbai* by Narayan Surne (translated into English by Mangesh Kulkarni, Jatin Wangle and Abhay Sardesai) is taken from Sujata Patel and Alice Thoner (eds.), *Bombay Mosaic of Modern Culture* (New Delhi: Oxford University Press, 1995), p. 147. The verse referring explicitly to Mumbai as a rotten culture is on p. 148, as follows: "We move on again, / settle in another vacant lot; / And live out the legacy / of this putrid culture".

The *dabbawalas'* customers can therefore safeguard their own food diversity, as they are not restricted to the limited choice of food available at affordable prices near their place of work, and at the same time they can preserve their own notions of food purity. Although the food they get delivered to them by the *dabbawalas* may fail to meet caste requirements in full, it usually does comply with most ethnic requisites, at least in the terms of ethnicity demarcation explained by Michael M. J. Fischer: "ethnicity is a process of inter-reference between two or more cultural traditions, and [...] these dynamics of intercultural knowledge provide reservoirs for renewing humane values. Ethnic memory is thus, or ought to be, future, not past, oriented".[114]

A Mumbai-style interpretation of Claude Lévi-Strauss's culinary triangle

When Claude Lévi-Strauss proposed his famous "culinary triangle" —borrowing his approach from Jakobson's concept of the "vowel triangle" as a basic semantic field offsetting universal principles of human linguistics —he explored the opposition of "elaborated" versus "unelaborated" food.[115] He set out to explain how a similar contrast among basic categories could serve as a model for the development of food preparation, cooking being, like language, a truly universal marker of humanity. He proposed to contrast the categories of "raw", "cooked" ("roasted/boiled") and "rotten" food as basic semantic markers of the complex interactions that transform raw food into cooked food (an essentially "cultural" process) and into rotten food (an apparently "natural" process). A further difference is drawn between roasting and boiling, which entails using a cultural object (a receptacle) to transform the food: "in as much as culture is therefore conceived as a mediation of the relations between man and the world, and boiling demands a mediation (by water) of the relation between food and fire which is absent in roasting". As "cooking effects [...] a mediation between [...] the burnt world and the

---

114 Michael M. J. Fischer, "Ethnicity and the Post-Modern Arts of Memory", in *Writing Culture: The Poetics and Politics of Ethnography*, ed. by James Clifford and George E. Marcus (Berkeley: University of California Press, 1984), pp. 194–233 (p. 201).

115 Claude Lévi-Strauss, *L'Origine des manières de table* (Paris: Plon, 1968). The Jakobson's concept is explained in A. Duranti, *Antropologia del linguaggio* (Rome: Meltemi, 2000); Lévi-Strauss's appraisal of his theory is briefly described in Claude Lévi-Strauss, "The Culinary Triangle", in *Food and Culture: A Reader*, ed. by Carole Counihan and Penny Van Esterik (New York: Routledge, 1997), pp. 36–43 (pp. 36–37).

rotten world", it can be viewed as a metaphor of cultural development.[116] Stretching this metaphor a bit, we can represent the *dabbawalas* as mediators between the strong undertow of a "rotten" city—one where ever shifting patterns of human interaction among individuals and groups with different backgrounds and often conflicting agendas constantly produce change even beyond the subjects' own reckoning—and the desire to somehow keep things under control. As agents of cultural "elaboration", enabling customers both to preserve culinary traditions and to facilitate their transformation, the *dabbawalas* inhabit the highly charged space between the culinary triangle polarities of the "raw" and the "rotten".

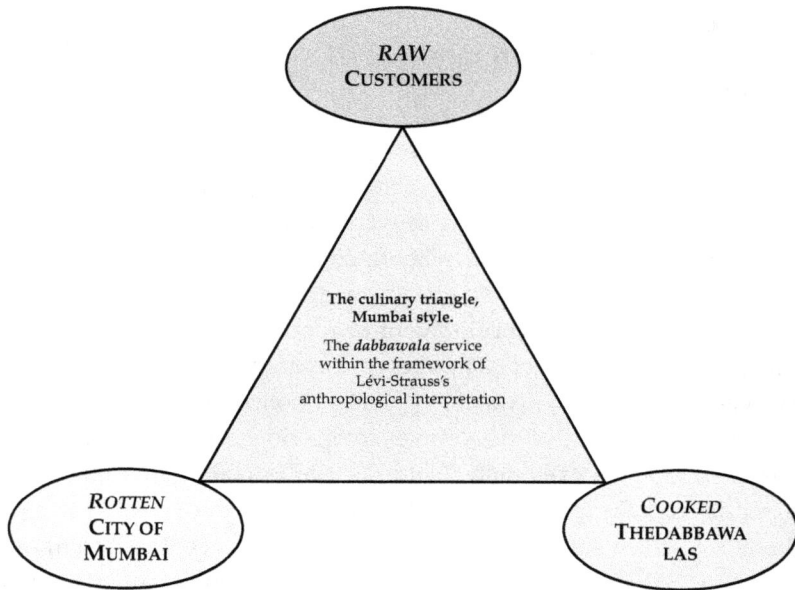

Figure 6.   The culinary triangle, Mumbai style. From Sara Roncaglia, *Nutrire la città* (Milan: Bruno Mondadori, 2010), p. 148. By kind permission of Bruno Mondadori.

In the Mumbai-style culinary triangle, new customers of the *dabbawala* system can be symbolically portrayed as being still "raw" to the city, i.e. untainted, unelaborated, yet ready to start their trajectories of cultural

---

116   Lévi-Strauss (1968), p. 426; Counihan and Penny Van Esterik (1997), pp. 36–37.

transformation. The *dabbawala* association inhabits the category of the "cooked", as they process customers' requests and transform their home-cooked food in a cultural expression of the city itself, thanks to the managerial skills and spiritual beliefs that uphold their tradition of service. The city of Bombay-Mumbai as breeding ground for "natural", unwarranted transformation of any culinary tradition that enters its premises may well be portrayed as the "rotten" pole. The city's many different souls absorb and develop cultural processes that the metropolis slowly makes its own, re-shaping and perpetuating human knowledge as cultures blend and feed off each other in this fertile "rotten" soil.

As a semantic system of oppositions, says Lévi-Strauss, the culinary triangle may serve "as a formal framework to express other oppositions, either cosmological or sociological".[117] In an ethnographic context, it can be usefully employed to clarify the transformations that take place at a social and cultural level. The *dabbawala* organisation has been built upon its members' devotion to the principles of Varkari Sampradaya, and upon a code of conduct that has been laid out according to their founder's intention to make their work the embodiment of those principles. Shaping their work into a form of spiritual practice, they turn a service economy into a more complex, almost alchemic process, one that infuses business transactions with new and deeper meanings. Just as cooking turns raw food into "socialised" food, so does the *dabbawalas'* work enhance the sociability of the city's inhabitants' eating practices, as their home-cooked food is often eaten among strangers.

If we take into account the spiritual sphere of Indian food, a parallel can also be drawn between the culinary triangle's three categories and the three qualities of food according to the Ayurvedic tradition. Thus, "raw" can be matched with the pure, *sattvic* food, while "cooked" can be linked to *rajasic*, "rotten" to *tamasic* food. The longing for *sattvic* purity that prompts at least some Mumbai customers to choose home-cooked, safe and "holy" food over the (mostly) *tamasic* food on offer in the city, can be satisfied by the purveyors of the *dabba* from the home to the workplace, performing a task that can ultimately lead to subtle adaptations in food preparation, gradually changing *sattvic* sensibilities into a willingness to adapt and experiment with the *rajasic*.

---

117   Ibid., p. 479.

This is how the *dabbawalas* can become middlemen capable of transporting "cooked" food to "raw" customers in a city that has a "natural" penchant for accelerating transformation thanks to its mixture of cultures and lifestyles: it is in this sense only that we can portray the process as a progressive "decay of purity". The importance given to the local "food code",[118] and to how it defines *rasa*, or "taste", in interpreting Mumbai, is grounded in the way this code appears to be dominating all others in defining social and personal identities, and—more generally—in outlining the human condition in a given social environment. Human beings experience social relations, place and space through cognition, based on the way they organise sensory perceptions.[119] This sensory input is processed within each culture in a specific way: as David Le Breton writes, "faced with the multitude of sensations possible at any time, any society establishes its own selection criteria".[120] To be part of a specific society also means acquiring a particular way of enjoying it. Though it proceeds from a common physiology and shared biological needs, the way a community codifies its perceptions and tastes results in a specific form of social organisation that provides guidance and structure to people's conceits and purposes.[121]

The appreciation for the Mumbaite food code's wide range of expression, with its thick cultural, religious and ethnic implications, is key to the *dabbawala* enterprise's enduring success. Beyond the diverse roles that food plays in the urban context and the vast variety of available choices, the popularity of their delivery service indicates that eating home-cooked food—i.e. eating "familiar" food—is a way to comply with rituals. Even in the city's context of unceasing and extensive transformation, the *dabbawalas* make it possible to sanctify every meal, preserving a body of knowledge that people are loath to lose. It is not just a matter of compliance with traditional and religious norms, but rather of upholding beliefs and understandings concerning food that transcends hedonistic considerations. Indeed, the *dabbawalas* concur in structuring a social order capable of accommodating the coexistence of different communities that, while living in close contact with each other, retain distinctive *corpora* of myths and rituals.

---

118   I use the term to translate *code alimentaire* and *code gustatif*, according to the definition that Lévi-Strauss gives in *Le cru et le cui* (Paris: Plon, 1964), p. 265.
119   David Le Breton, *La Saveur du monde: une anthropologie des sens* (Paris: Editions Métailié, 2006), p. xiv.
120   Ibid.
121   Ibid.; and Paul Stoller, *The Taste of Ethnographic Things: The Senses in Anthropology* (Philadelphia: University of Pennsylvania Press, 1992).

The intricate clockwork by which Mumbai society perpetuates itself in daily life is mirrored by the way the *dabbawala* delivery system incorporates the urbanised population immigrant experience.[122] The complex architecture of spirituality, service ethics, respectful preservation of food requirements (and therefore of the food's ingredients, as well as of the ways in which food is prepared and processed), and—above all—the social ties subsumed in the preparation and consumption of meals all contribute to this metropolitan immigrant society's symbolic sphere.

The *dabbawala* experience can thus also serve as a lens to focus our understanding of the deep emotional and spiritual dimensions that contribute to the city's cosmology, one that can be shared by all souls that have chosen to live in it. A subtle alchemy of tastes that is also capable of interacting in surprisingly vibrant ways with the stimuli provided by a global economy of which the city of Mumbai is itself the offspring. As a case study, the *dabbawala* system shows how a form of specialised, culturally-embedded service economy, can empower itself by accessing the wealth of history and tradition of its territory, and can subsequently develop a strong competitive advantage.[123] The *dabbawalas* achieve this commercial success without betraying the heterogeneity of their own cultural roots. Indeed, they succeed by making the intricate mixtures not a factor of chaos but a dynamic form of order.[124]

---

122   I specifically refer to Abdelmalek Sayad's reflection on the capacity of the migrant's construction of social experience to act as mirror of symbolic structures and required practices that characterize the host society, see Abdelmalek Sayad, *L'immigration ou le paradoxe de l'altèrité* (Brussels: De Boeck Université, 1991).

123   Nicola Bigi, "Quell'atmosfera culturale che rende economicamente speciale una città" (interview with A. J. Scott) in *Dialoghi Internazionali* 6 (Milan: Bruno Mondadori, 2007).

124   Jane Jacobs, *The Death and Life of Great American Cities* (New York: Random House, 1961).

3. Nutan Mumbai Tiffin Box Suppliers Charity Trust: The Shaping of *Dabbawala* Relations

After the initial rudimentary cooperative was set up in 1954, the Nutan Mumbai Tiffin Box Suppliers Charity Trust (NMTBSCT) was registered with that name in 1984. The name was developed to include all the elements that characterise the work of the *dabbawalas*: the city where tiffin delivery is offered; the organisation's specific role as a distribution network; and its establishment as a charity trust, reflecting its social commitment to sponsoring various non-profit projects.[1]

An executive committee of thirteen permanent members sits at the highest level of the NMTBSCT, and it is responsible for defining and fine-tuning the overall *dabba* transport system in Mumbai. A second tier consists of about 800 *mukadams*, who are the group leaders in charge of a team of five or ten *dabbawalas*. The rest of the organisation is made up of the *dabbawalas* themselves, the members of the association. In the words of NMTBSCT's president, Raghunath Medge:

> Ours is an association, a union of 5,000 people. There are many groups in a single association and I'm the president of the association representing all the Bombay groups. I'm also chairman of my group, so I'm given a salary. The association's work is a social commitment, which entitles it to qualify as a charity trust. Each station has at least five or six groups. Gangaram Talekar is the secretary general and there are nine directors throughout Bombay, found in the various areas. Then there are the *mukadams*, who are the heads

---

1   All information contained in this chapter is based on interviews, personal communications and fieldwork undertaken in Mumbai in 2007. For further details of this fieldwork, please refer to the Appendix.

DOI: 10.11647/OBP.0031.03

of the working groups, and the 5,000 members, who are not employees but partners, members. There is competition among the groups: we compete to find new customers, just like anyone else.

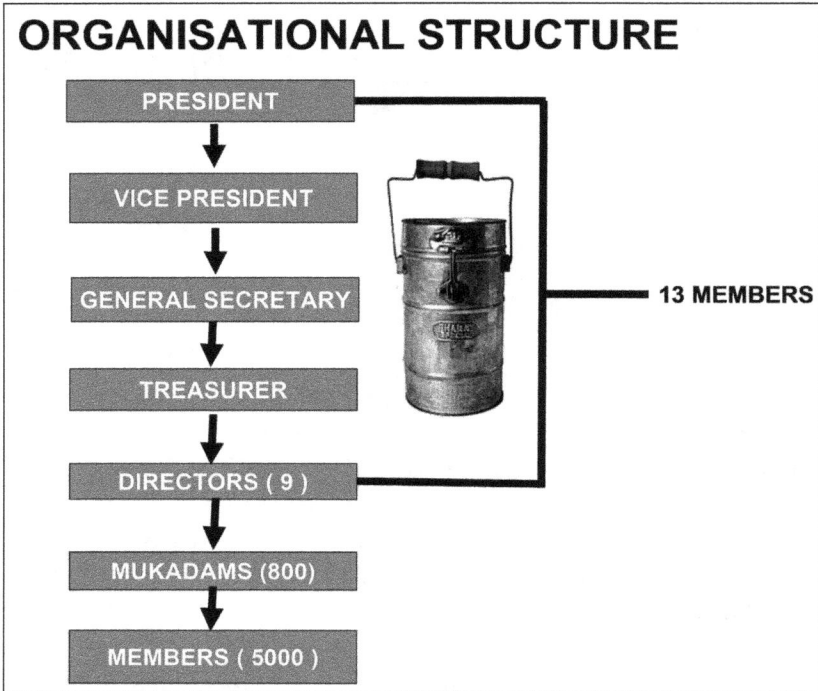

ORGANISATIONAL STRUCTURE

PRESIDENT

↓

VICE PRESIDENT

↓

GENERAL SECRETARY

↓

TREASURER

↓

DIRECTORS (9)

↓

MUKADAMS (800)

↓

MEMBERS (5000)

13 MEMBERS

Figure 7. Diagram by Pawan G. Agrawal, director of Mumbai's Agrawal Institute of Management. By kind permission of Raghunath Medge.

The executive committee

The executive committee is elected every five years and comprises a president, vice-president, secretary general, treasurer and directors. Committee members meet every month to discuss any problems related to the service and the association's line of operations. Crucially, all the people who hold these offices continue to operate as *dabbawalas* because their salary comes from distribution work. Even the president does not draw a salary on the basis of his rank but on his *dabba* delivery line.

As president, Medge is one of the pillars of the NMTBSCT. He combines brilliant communication skills with the ability to transmit a vision of shared values.[2] The sense of community in the association derives from the shared

2   Medge has become a great spokesperson for the *dabbawalas*. After his first meeting

cultural background of the *dabbawalas*, but also from each *dabbawala*'s awareness that he is "part of an important project that generates meaning", an awareness reinforced by the executive committee and by the president.[3] There is a consolidated tradition that Medge has exploited whilst acting as president to leverage the *dabbawala*'s conceptual models—i.e. the images and figures that influence how a *dabbawala* interprets the world and consequently how he acts—to achieve a mutual objective and working trust amongst the members.[4] This does not mean there are no conflicts amongst different groups, but these differences always play out within an association working for a common purpose. Social interaction between the *dabbawalas* is encoded within a corporate culture that uses a "policy of emotion management" to create a shared work ethic and enable socialisation which, especially in a migrant context, helps to overcome moments of loneliness by sharing holidays, free Sundays, and times of "sorrow and joy".[5] Medge explains:

> Not just anyone can be the president, it has to be a *dabbawala*. At the moment, I have the Vile Parle contract. If I wasn't a *dabbawala* I couldn't be the president. I'm responsible for the whole Bombay group. To be president you have to have specific characteristics, you need to know things, you need to know

---

with Charles, the Prince of Wales, in Mumbai in 2003, he was invited to the Prince's wedding to Camilla Parker Bowles in 2005. These two events aroused the interest of the international press and gave the *dabbawalas* unprecedented media popularity. Studies by Indian and foreign researchers on the *dabbawala* style of management rocketed. The association's official website offers articles to download, a direct line to the president, as well as an option for spending a day delivering *dabbas* with a group. See http://www.mydabbawala.com/general/aboutdabbawala.htm [accessed 30 January 2012].

3   Augusto Carena, "Un'organizzazione che apprende", in *Io erano anni che aspettavo. Impresa, lavoro e cultura in uno stabilimento del mezzogiorno d'Italia: la Barilla di Melfi*, ed. by Giulio Sapelli (Parma: Barilla 2009), p. 41. Research was undertaken in the Melfi plant with the participation of Augusto Carena, Roberta Garruccio, Germano Maifreda, Sara Roncaglia, Veronica Ronchi, Giulio Sapelli, Andrea Strambio and Sara Talli Nencioni. Carena's work draws upon a systemic reading of the Melfi plant and takes Peter Senge's "Fifth Discipline" as the template for the structure, setting up five interdependent pivots: systems thinking, personal mastery, mental models, group learning, and team learning. See Peter M. Senge, *The Fifth Discipline: The Art & Practice of Learning Organization* (Danvers, MA: Doubleday, 1990).

4   Carena (2009), p. 44.

5   Catherine A. Lutz and Lila Abu-Lughod (eds.), *Language and the Politics of Emotion* (Cambridge: Cambridge University Press, 1990). Much has been written on the concept of "corporate culture". Amongst others, see Edgar H. Schein, *The Corporate Culture Survival Guide* (San Francisco: Jossey-Bass, 1999); Geert Hofstede and Gert J. Hofstede, *Cultures and Organizations: Software of the Mind*, 2nd ed. (New York: McGraw-Hill, 2005). In this case I mean business culture as the intersection of different cultures or micro-cultures within the association. Each group has a strategy for developing its task based on a shared framework, but also on the possibility of managing that task independently. The micro-cultures may not be in harmony with each other and their interactions may result in conflicts and contradictions. This is because the company or the trust, belonging as it does to the social and symbolic order in which it is located, is a sphere that interacts with it and is therefore susceptible to the same processes of conflict and solidarity.

about the world, about Bombay, the problems arising in tiffin work. The president is responsible for keeping the group together. How can anyone who fails to keep his family together keep 5,000 members together? To see if someone will be a good president, we test his strength, his intelligence, his knowledge, his qualifications, how he bonds with the group, how he interacts, if he is able to talk straight with everyone. Sometimes he has to be gruff and to do that he has to be confident and show it. He has to be able to control the whole group, whether he knows how to do it or not.

A short story from an NMTBSCT dabbawala: free time and faith

I came from the village of Rajgurunagar, near Pune, and I've been in this family [in other words working as a *dabbawala*] for ten years. My own family is all back at the village. I'm on my own here. Someone from the village brought me here to work as a dabbawala. In Bombay I live near Four Bangla, at Sagar Kuti, where we have a room. My work starts at Four Bangla and goes as far as Vile Parle. We are an association of people working as *dabbawalas*. We meet after work and we pass the time performing plays. We go to *ramlila* [a pilgrimage] and sometimes there's Pandhur [this may be the festival of Pandharpur, which is the sacred place of the Varkari], we play the *lezid* [harmonium]. We go on the *ramlila pandhupratha* [pilgrimage] to the temple of Ramgir Baba Garon. We also play the harmonium and I play the *taal* [a kind of drum also called a *dholak*]. But we stopped playing in Bombay about five years ago: we do it at the village.

On Sundays we stay at home with our families. Anyone who doesn't have a family here and is alone goes to their friends. You know yourself how nice feast days are. Anyone who wants can go, because in this country we are free. Anyone who wants can take Sunday off and go to their family or their children, or do errands they can't do during the week, or meet their friends, meet people. Anything you can't do from Monday to Saturday, you can do on a Sunday. We *dabbawalas* meet for the festivities and celebrate them together. For instance, Holi [the festival of lights] or feast days on the Hindu calendar. For Holi we all meet the day before, because we don't work on Holi itself. During Holi we play with colours. We play where we work but if we go to the village, then we really celebrate Holi properly, better than in Bombay. They don't do so much in Bombay. We share every joy and every sorrow. It doesn't feel as if we've gone so far away: the village is close. All my family is there:

my wife, my sons and daughters. Two of my children are married and two still go to school. One is in seventh class, the other is the fourth class of the Marathi school [taught in Marathi]. One is twelve and one is eight. I'm Marathi, they are *Varkari*. [He shows his necklace] This is a Pandharpur necklace. As Medge says: 'Work is worship', no doubt about it. We all believe that.

Each area of Mumbai served by the *dabbawalas* has a director and the areas are defined in relation to a railway station. One of these is Borivali; others are Kandivali, Malad, Andheri and Thana. The directors are in contact with one another and ensure there are no problems in distribution. For this reason, they do not have an office but work on trains, station platforms, *dabba* handover areas and places where organisational issues may arise. The job of the director is appointed based on skills and seniority, since the delivery work is physically very tiring. Aspiring directors must also have an aptitude for managing human resources, which can be seen in their vision of their work group as a real family and in their loyalty to the association's values, expressed in concrete terms by worshipping Varkari Sampradaya. Medge describes the role of the director:

> The directors are not in an office and I also go out with them, wandering around the trains and platforms. After lunch, after tiffins have been delivered, everyone is free and can relax. Then we all go to lunch at Grant Road. Then, after 4 pm, we go to Andheri, where we have one of our association offices.[6] After 7 pm I go home. The director considers the group to be a family where everyone works, teaching that along with the gift of food there is the gift of knowledge. In this spirit we also give presentations in colleges because only the gift of knowledge is greater than the gift of food.

Gangaram Talekar, the NMTBSCT secretary, adds:

> You move to management when you can't do heavy work or when you get old. To be a director you have to be involved [in the association for several years]. And you must be able to do other jobs like this. The association started appointing directors in 1992, usually electing a director from one of the workgroups for a term of five years. Mine was a typical evolution, for instance first I was a *dabbawala*, then a *mukadam*, then a director, then treasurer. Now I'm the secretary and I'll be in office for five years.

---

6   The NMTBSCT has three Mumbai branches: Dadar, Andheri (E) and Grant Road.

A short story from the NMTBSCT director

I've been in this job for twenty years. My father worked in the fields. When I arrived in Bombay I didn't join the *dabbawalas* immediately. First I went to Malad, because my sister was there and I worked with her for five years, then I started this job. I rented a room at Andheri, in Sher-E-Panjab, and I became a *dabbawala*. Many of the other villagers were *dabbawalas* and I liked this job too, so I started with the basic salary. After eight years I became a *mukadam*. I gave money to my *mukadam* a little time at a time. I saved up and slowly paid for the line. Now I have two lines in Andheri and I'm both a *mukadam* and a director. The roles are similar, so if someone working in my group doesn't turn up, if they're ill or they go back to the village, then I do their job. It used to be very good in Bombay, with not too many people. Here in the village there wasn't much money and getting to the city was cheap. The work was agreeable and the trains weren't jam-packed. It was fine but now it's too crowded. The crowds have increased as we've got older. I'm 50, so I no longer enjoy this work. The weather in my village is good and it's a good place to be. Bombay's too hot, there is a lot of pressure. I used to like it; when you're young it's different. The younger ones enjoy it but the older ones don't.

The second line of operations

The second line of operations is coordinated by over 800 *mukadams*, who supervise the tiffin route as far as the final delivery.[7] The *mukadam* participates in the recruitment of new *dabbawalas*, assessing their suitability by taking into consideration both their reputation and their shared origins with other members of the association. He also manages relations with customers, making preliminary agreements for deliveries, and administers monthly subscriptions (at a cost of about 120 rupees per month).

As can be seen from the twelve points in the agreement form (Figure 8) the delivery system is based on a code whose observance is enforced by the *mukadam*, who also oversees any disputes that may arise among various

---

7   Traditionally the term *muqaddam* (literally "village chief") indicated the upper peasant class—a sort of rural aristocracy who did not farm the land directly but enjoyed special privileges over the land and crops. See Michelguglielmo Torri, *Storia dell'India* (Bari: Laterza, 2000), p. 194.

dabbawala groups as well as having the more difficult task of enhancing network competitiveness to improve earnings.

॥ श्री ज्ञानदेव प्रसन्न ॥

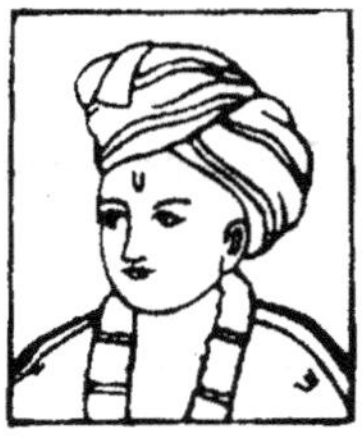

फोन: दादर ४२२१८२७
ग्रँट रोड ३८६०७४२
अंधेरी ८३७१८९७

टिफीन, कॅरीअर मुकादम

धोंडु कोंडाजी चौधरी

२७२, लक्ष्मीनगर झोपडपट्टी, लिंकींग रोड,
गोरेगाव (प), मुंबई-४०० ०९०.

नं : 5970

चुक भूल दयावी घ्यावी. तारीख / / २०० माहे २००
मध्ये भोजनाचा डबा पोहचविण्याची आम्ही हमी घेतो. वसुल
करणाऱ्याची सही. धो. को. चौधरी

1) One months Salery is Diwali Bonus. 2) Please do not pay without Bill. 3) Please do not keep silver or stainless steel utensils in carriers. 5) If carriers are lost only value will be paid in the monthly payment of half. 6) We are not responsible for carriers lost in your office. 7) Delivery will dopend as per train timings. 8) If trains are last for 1-30 Hours of more the carrier will be return as it. 9) If you have any complaint about oure servents Please let us know by post. 10) Please look carrer having single supporters hance we will not be held responsible for damage of the lunch. 11) Payment to be made in fulll even carrier fails to carry for 8 to 10 days in a month 12) If carriers lost we pay half from my only.

Figure 8. Client agreement form. By kind permission of Raghunath Medge.

Each of the approximately 120 groups present on Mumbai territory is independent of all the others: it is a "Strategic Business Unit" and has to increase its customer base in order to generate the *dabbawalas'* wages.[8] Medge uses the metaphor of a cricket team to describe the *mukadam*'s role:

> The *mukadam* supervises the group, is an expert in everything that concerns the work, is 'the reserve'. For example there are fifteen people in a cricket

8 Shrinivas Pandit, *Dabawalas* (New Delhi: Tata McGraw-Hill, 2007).

team, but only eleven are fielded: the others are the reserves. That's the *mukadam*, he's the reserve. He keeps the group together, he's the player who knows how to do everything. He holds them together, manages customers, helps the *dabbawalas* load and unload. The trains stop for deliveries for ten or fifteen seconds, so he has to help. His work is very important because he has to manage lots of people and it's up to him to be able to coordinate the group, to look after the *dabbawalas*, increase revenue. The *mukadam* is like the captain of a cricket team: if the team captain is good, then the team will be a success.

A short story from an NMTBSCT mukadam

The name of my village is Rajgurunagar and it's in the Pune district. I worked in the fields there. There are five of us altogether: four are in Bombay and one is in the village, still working in the fields. I came to Bombay in 1960 and started with tiffin work in 1967. I used to work in Dadar but now I work at Andheri and I'm a *mukadam*. My father wasn't a *dabbawala*, he worked in Bombay near Victoria Terminal. He loaded and unloaded many of the ships that docked. I've been in this job for seventeen years and I've been a *mukadam* from the start. I have ten to twelve people working with me because I bought the line [the line is the specific route assigned to a *mukadam*]. There are forty tiffin in a line [here the term line is used to indicate the basket for carrying *dabbas*], which I bought from another *dabbawala*. Now my Andheri group has about thirty to thirty-five lines. A *mukadam*'s work includes filling in for a man if he doesn't come to work. For example, if a *dabbawala* doesn't come to work today, perhaps he's sick, so I'll work in his place. If there are problems at the station, they call me: if they need help to load trains, or if there are problems with traffic, or parking. But there aren't any big problems.

The *dabbawalas*, the *mukadams* and the thirteen figures that make up the executive committee are members, "freelancers": no one is an employee. The association has no institutional hierarchy but an agreement for decentralised operations, with each group using its own resources to extend the customer base. The current contractual formula was drawn up in 1982, after two incidents (the 1975 railway worker strike and the 1982 cotton worker general strike) induced the NMTBSCT president to modify the association's statute, changing it to a cooperative. Gangaram Talekar explains:

> When I started as a *dabbawala* I was a *mukadam*. There weren't many *mukadams* then but lots of employees. Each *mukadam* had eighteen to nineteen people working for him. Until 1980 everyone was an employee, including the *mukadams*, and everyone was equal. After 1980, they all became members. If someone wants to go to the village, we need men to replace him. Customers don't stop eating, they have to eat. Even if a *dabbawala* doesn't work, customers still have to have lunch. So they might buy food from outside, but then we lose the customer and work. Before, *dabbawalas* had a salary and as such they worked as employees, so work went like this: today they turn up, fine; tomorrow they say they're feeling poorly and don't show up. So if they're all members, things are different.

A *dabbawala's* earnings derive in part from the ability of each group to attract more customers into their network but also partly from the role played by an individual *dabbawala* in that network. Those who have just joined the association and have not purchased a *mukadam* line have a fixed basic wage that may vary according to the group's revenue (from 2,500 to about 4,000 rupees a month). A *mukadam* supervising a group earns on the basis of how many tiffins his men can deliver; if he supervises various groups, who deliver tiffins in various parts of the territory, his earnings may be higher (about 6,000–7,000 rupees a month). To become a *mukadam*, an ordinary *dabbawala* has to buy a customer line being auctioned off, which occurs when a *mukadam* decides to retire and sells off their lines because their children are not interested in taking it over. One NMTBSCT *dabbawala* describes how he began:

> I've been doing this job for six years and I come from Pune, from a village near Var. I came to Bombay because at home we had problems with the rain. Without rain there's nothing at the village. After working so hard here, I now know what it's like in Bombay, with crowds and traffic. Not all *dabbawalas* are members. The ones in the network are members. The ones who aren't are not members. To be a member you have to register with the association and pay twenty-five to thirty rupees in dues. When I started working I wasn't a member, I became one when I bought a tiffin line. You have to get a line. For example, if someone has a line that costs 5,000 rupees and he has to go back to the village because he's tired or because he has to do another job, then he says: 'I have to sell'. If he wants to sell then he auctions to the highest bidder. For example, for one rupee I give five; for another I give eight, for yet another I give ten, then depending on the value of the line, a bid is made. I bought mine for 40,000 rupees.

Another *dabbawala* explains how the system works:

> I'm a member too and fifteen years ago I bought the line for 30,000 rupees. The line had forty tiffins. The price of the line is not based on the number of tiffins, but the length of the route you cover to deliver them. For example, ten tiffins can be worth 2,000 rupees a month but so can two tiffins. The price of tiffins is different and this means the price of the line is different. So, a 2,000-rupee line may cost between 8,000 and 16,000 rupees. If a line is worth 20,000 rupees, we'll pay 80,000 rupees for it. We make a bid based on a calculation of how much the line earns every month. Anyone who doesn't have the money can't become a member so they work for a wage: basic pay is 2,000, 2,500, 3,000 rupees. If a *dabbawala* closes his business, no one can deliver tiffin in his place because if someone else goes, people complain to the office. If a customer complains 'I don't want this *dabbawala*, send another one!' the office can't do anything about a *dabbawala*. But I'm a member and if someone delivers tiffin in my place, then I can complain about it to the office: 'This man is taking my tiffin to deliver!' So people from the office will tell him: 'Don't take his *dabba*' and the office also fines people. That's why we all work well.

Data supplied by Medge made it possible to reconstruct the costs of managing and the net earnings of an individual *dabbawala* member.

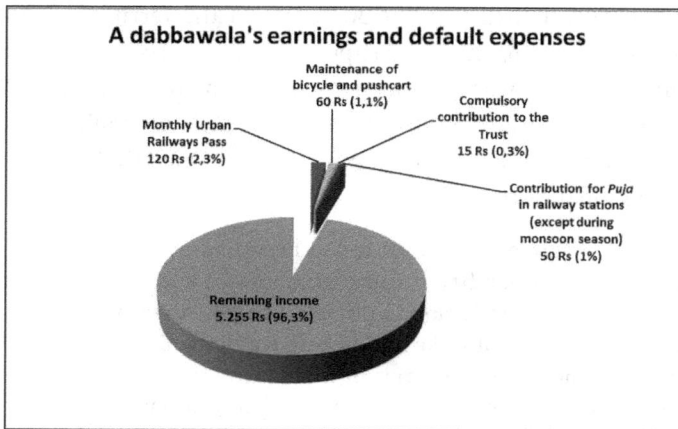

Figure 9. *Dabbawala* Costs Managing.[9]

---

9   Data: Ramasastry Chandrasekhar, *Dabbawallahs of Mumbai*, Richard Ivey School of Business, University of Western Ontario, 2004, available at http://beedie.sfu.ca/files/PDF/mba-new-student-portal/2011/MBA/Dabbawallahs_of_Mumbai_(A).pdf [accessed 28 October 2012].

Distribution logic[10]

> We don't cook. The cooking's done at home. We're not a catering service, we're just a [distribution] network. You have to imagine food cooked in each customer's home and delivered to their place of work. That is our service. Caterers cook food. If no one at home can cook the food, then I give the phone number of a caterer. 'I don't have anyone at home to do the cooking, so can you cook for me?' We have the numbers of caterers, so if you have problems at home and there's nobody there, or they don't have time to give children food, then we can give a number. If that's good with the customer, they settle a price for the delivery and the cooking service. But not us, we're just distribution.
>
> — NMTBSCT *dabbawala*

The NMTBSCT has the monopoly over the meal delivery service and, thanks to both its excellent grasp of distribution logistics and the high reputation it enjoys as an *annadatta* (food deliverer), it has gained a considerable competitive edge in a specific segment of the Mumbai market. Traditionally, logistics were regarded only as a function that allowed the enterprise to optimise materials, goods and intangible flows, like information. With the emergence of the "supply chain" concept that the *dabbawalas* appear to use in their management approach, the whole logistics process has been redefined to optimise links and coordination among suppliers, customers and distribution.[11] In this respect, the inventory levels and flow of goods in the supply chain have been optimised, with an increase in the production efficiency of the enterprise and its fulfilment of incoming orders, while improving customer service by keeping prices down. The association therefore appears as an organisation capable of planning, implementing, and monitoring delivery operations, and as an expert in the "art" of moving materials, people and information from one place to another in order to satisfy customers.[12]

What is known today as "supply chain management" has been effectively, albeit more or less unintentionally, internalised by the entrepreneurs of Mumbai's meal delivery sector. The logistics essential to managing the distribution network revolve around the availability of urban

---

10   This section was written on the basis of my observation and participation in the delivery service.

11   Nasreen Taher, *Impeccable Logistics and Supply Chain Management: A Case of Mumbai Dabbawallahs* (Hyderabad: Icfai University Press, 2007). For a review of "supply chain management" practices and theory, see Claudio Ferrozzi and Roy Shapiro, *Dalla logistica al Supply Chain Management. Teorie ed esperienze* (Turin: Isedi, 2000).

12   C. S. Parekh, *The Dabbawallas of Mumbai* (unpublished PhD thesis, Narsee Monjee College of Commerce and Economics, Mumbai, 2005).

infrastructures and the cultural approach that *dabbawala* customers have to food. The *dabbawalas* adapt their distribution logistics and planning process to customer needs, taking into account the flexibility of their own working group. In 1998, the American magazine *Forbes* conducted a study of the *dabbawala* service and awarded its organisation a 6 Sigma, with a 99.9999% accuracy rate.[13] This means that only one tiffin in every six million deliveries goes astray.[14] Natarajan Balakrishnan and Chung-Piaw Teo, researchers at the National University of Singapore, compared the *dabbawala* distribution system with that of postal delivery and of a Mumbai goods retailer.[15] In the first case, the mail is sent to a single central sorting branch and then delivered to the final recipient via a hub that handles distribution operations. The goods retailer, on the other hand, uses a zone map system similar to that of the *dabbawalas*, i.e. identifying groups of vendors within coded zones and then supplying each of these through a sub-sorting unit.

Although logistical considerations are important in meal delivery organisation, there is a strong likelihood that the *dabbawala* system relies mainly on Mumbai's specific culture which, in turn, orients the executive committee's planning. The *dabbawalas* identified Mumbai as a source of opportunity and their delivery process developed by taking into account the changing metropolis and the evolution of the preferences and well-being of the inhabitants, the urban infrastructure and social characteristics. Indeed, the service is difficult to replicate in other cities precisely because several elements characteristic of Mumbai are absent in other urban contexts—a very extensive transport network and large working class, combined with the cultural unity of the *dabbawala* association rooted in the rural areas around the city. Considering these aspects, it can safely be said that Mumbai is the cultural milieu underpinning the *dabbawala* distribution rationale, the mental map that underlies their work and from which they draw inspiration.

The delivery process

The *dabbawala* starts work at about eight-thirty in the morning, when he cycles or walks to pick up *dabbas* from the door of the "customer-supplier",

---

13   S. Chakravarty, "Fast Food", *Forbes Global*, 8 October 1998.
14   U. K. Mallik and D. Mukherjee, "Sigma 6 Dabbawalas of Mumbai and their Operations Management: An Analysis", *The Management Accountant*, 42 (2007), 386–88.
15   See Natarajan Balakrishnan and Chung-Piaw Teo, "Mumbai Tiffin (dabba) Express", University of Singapore, 2004, available at http://www.bschool.nus.edu.sg/staff/bizteocp/dabba.pdf [accessed 17 July 2012].

usually whoever does the cooking. Time is of the essence in this process, because if one of the two parties is running late, the subsequent schedule fails. On average, each *dabbawala* is responsible for collecting thirty to thirty-five *dabbas*, the number depending on personal ability to memorise customer addresses and the physical strength for carrying the tiffin baskets. One NMTBSCT *dabbawala* describes his day, and the efforts made to deliver his tiffin on time:

> I pick up at least forty tiffin from homes just in Lokhandvala, from I, II, III and IV Street. Sometimes, when a bank is closed for holidays, there aren't so many *dabbas*, but we still have to go to the customer's home. I start work at 8.30 am and by 10.30 I have to be at the station, because my train leaves at 10.38. I cannot risk missing it, ever. Often someone doesn't have the *dabba* ready in time and it's late, but we adjust, we make it. For instance, we run faster up the stairs. If we know we're more than five minutes late, we pedal faster. And if we miss the usual train, we catch the fast one so we can make up ten or fifteen minutes. If we don't make the train, we use our lunch break to deliver tiffin and we eat afterwards. 99% of the time it doesn't happen but if it does, we skip our lunch so we can deliver food on time. Then we take back the *dabbas* and we go and sit in the station to eat. But in certain circumstances it's just not possible to deliver on time, for example during the summer monsoon when it rains, so sometimes we are gridlocked! When it rains, that does happen sometimes.

After this initial collection stage, the containers are taken to the nearest station by bicycle or in wooden baskets. Here a second group of *dabbawalas*, from the same line, takes the previously collected *dabbas* and loads them onto the trains. Although there is no formal agreement with the railways, the goods compartment at the head of the train is left for the *dabbawalas* or people carrying bulky goods. The biggest difficulties are the crowded stations and trains, which always make it problematic to move the heavy baskets among people trying to board the trains. *Dabbas* have to be loaded very quickly, in the thirty seconds the train stops on the platform. After this second stage, all the precious lunches are ready to move on to their destinations. If the trip is very long and includes a line change, the *dabbawala* in charge of the final delivery takes his own *dabbas* to a collection and sorting point. There are several strategic nodes near railway stations that serve as main centres for final sorting. In this case, the figure of the *mukadam* becomes essential for efficient coordination of the delivery network to ensure that no *dabba* is lost or routed to a wrong destination. The third stage is the final delivery: from the strategic collection point the cooked lunch is taken to the place of work of the "receiver-customer" for about 12.30 pm. The tension gradually eases and the *dabbawalas* can rest, eat their lunch and, lastly, prepare to make the journey back, following a circular route that begins and ends in the same way every day of the week except Sunday.

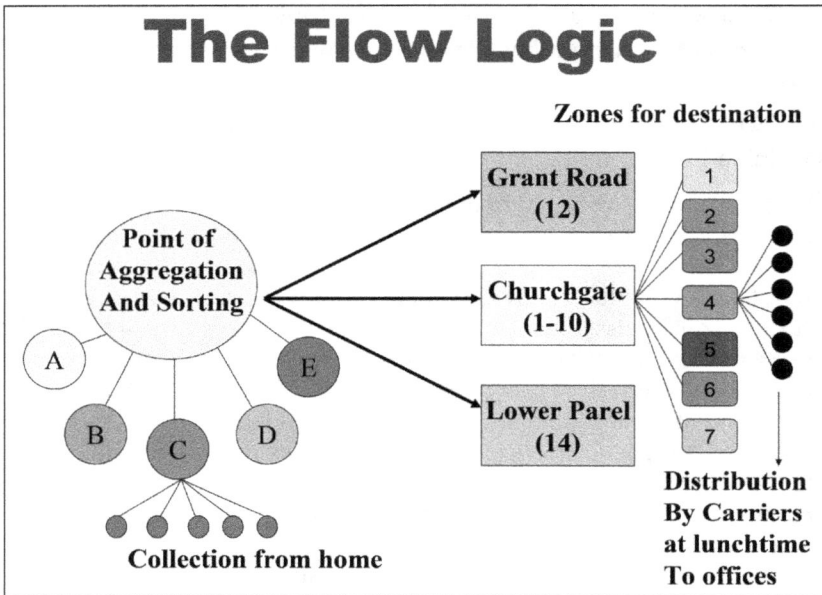

Figure 10. The Flow Logic. Diagram by Pawan G. Agrawal, Director of Mumbai Agrawal Institute of Management. By kind permission of Raghunath Medge.

The delivery process is easier to understand by looking at a schedule of the stages involved in *dabba* distribution:

8.20	The *dabba* is prepared by the "supplier-customer" and left outside the front door.
8.25	The *dabbawala* arrives and picks up the *dabba*. If he does not see it, he knocks on the front door.
8.35	The *dabbawala* loads the *dabba* into his tiffin basket or onto his bicycle along with others picked up in his area.
9.25	The *dabbawala* arrives at the *dabba* collection area of the nearest railway station.
9.30	The sorting process starts and *dabbas* are grouped according to where they have to be delivered.

9.40	When the train arrives, the *dabbawalas* board the compartment at the head of the convoy.
10.15	The train arrives at the major junctions and if the *dabbas* have to change line to reach their destination, they are delivered using a relay system involving another *dabbawala*.
11.00	The *dabbas* change trains and continue their journey.
11.45	Arrival at destination station.
12.15	The *dabbas* are loaded onto various baskets or bicycles and taken to the "receiver-customer".
12.30	The *dabbas* arrive at the place of work of the "receiver-customer".
Afternoon	The delivery process is reversed and the empty *dabba* is collected at about 1.30 pm from the "receiver-customer" and returned to the "supplier-customer".

Table 1. Schedule of the *dabba* distribution.

The *dabbawala* alphabet

A system like this could not exist without a code for identification of the *dabbas*. The containers change hands several times during the day, so the group must be able to recognise them or they may be lost along the way. Most *dabbawalas* are completely illiterate or barely able to read and write, so tiffin delivery relies on the use of identification systems to ensure successful delivery. These systems were an important factor in network development and basically comprise four or five symbols of different colours painted on the containers. Nevertheless, they do not share the same style due to the *dabbawala* association's characteristics, which gives each group the freedom to manage its work independently. Thus, the codes are often indigenous, derived from the regions of origin, or are symbols that bring to mind specific circumstances. In most cases, however, they are symbols common to the Indian cultural context, like letters of the Devanagari alphabet, religious allegories or just geometrical symbols, some of which are shown below:

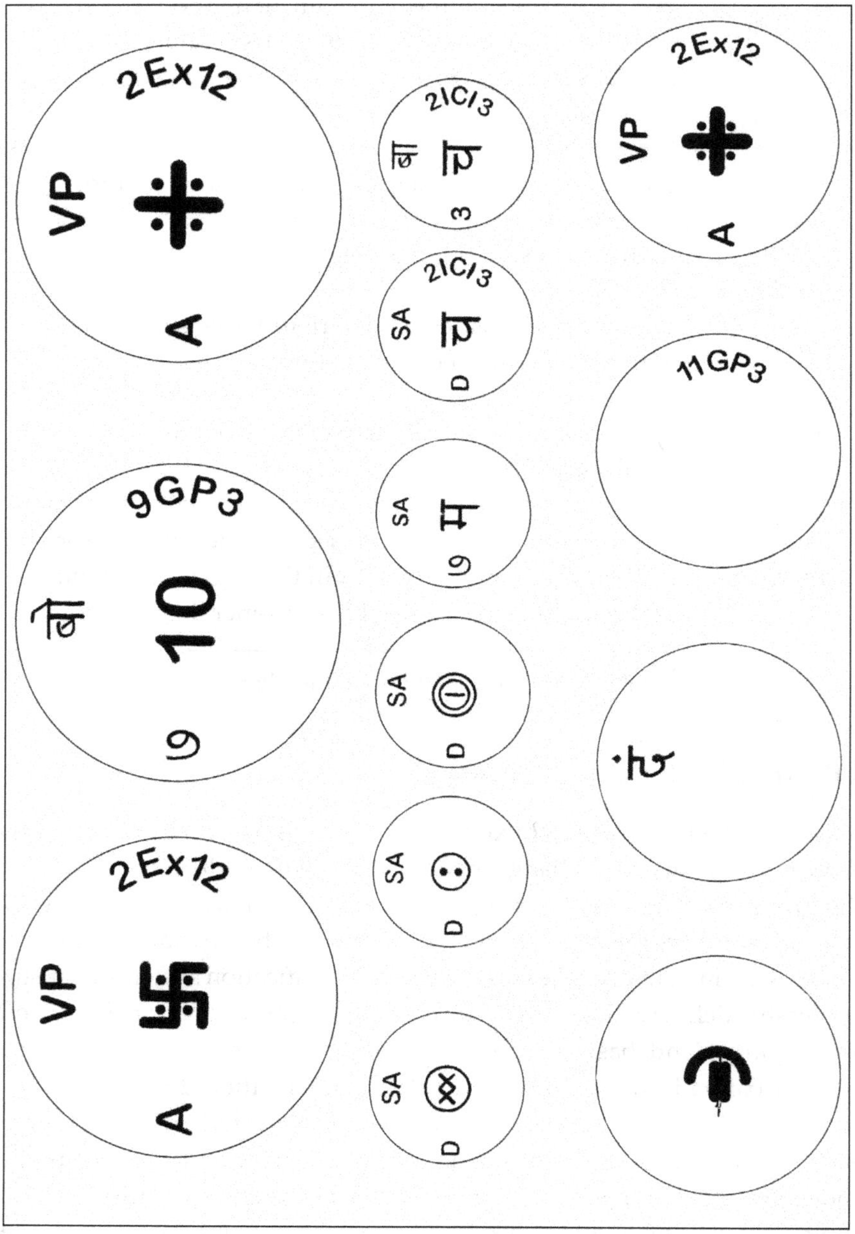

Figure 11. Examples of *dabba* symbols. By kind permission of Raghunath Medge.

GLOSSARY OF SYMBOLS[16]	
Dotted, right-facing swastika	*Swastika*: from the Sanskrit *su* "well" and *asti* "being". This is the most frequent representation of the swastika used in Hindu symbolism. It is an ancient symbol, often linked to the Sun and the solar cycle (the four arms may represent the seasons). It is an emblem of transformation, of infinite steps of status, of cyclical eternity and, as such, has assumed important connotations in various religions and philosophies. In Asia it is found in Hinduism, in Tibetan shamanism (*bon*), in Jainism, Buddhism, etc. In Europe it is seen in megalithic civilisations, Germanic and Celtic shamanism, etc. Often the orientation (right-facing or left-facing swastika) or the type of arm (curved or hooked swastika) will change according to the cultural context. The Hindu swastika is a positive symbol that symbolises the invigorating power of the Sun, which is renewed every day, hence the allusion to well-being. Traditionally the right-facing swastika is connected to Ganesha and is considered an auspicious symbol for the start or the inauguration of concepts or businesses, so it is seen at the entrance to buildings of worship, on book covers and at the front doors of homes. The left-facing swastika is consecrated to the goddess Kali.

Typically in India (and in Tibet, where this symbol is known as *norbu shi khyi*), the swastika is drawn with four dots between the arms, an allusion to the four *purusharths* or duties/ principles/achievements/blessings of the Hindu person: *dharma* (righteousness); *kama* (pleasure); *moksha* (knowledge); and *artha* (wealth). It is also read, above all in the *vajrayana* Buddhism, as a reference to the four "bodies" or *koshas*:

– *annamaya kosha*: the shell of food (physical body)

– *pranamaya kosha*: the shell of the vital force (five life breaths and internal organs)

– *manomaya kosha*: the shell of the mind (mind and perception)

– *vijnanamaya kosha*: the shell of intelligence/understanding (mind and senses). |

---

16   See H. Sarkar and B. M. Pande, *Symbols and Graphic Representations in Indian Inscriptions* (Delhi: Aryan, 1999); and Anna L. Dallapiccola, *Dictionary of Hindu Lore and Legend* (London: Thames and Hudson, 2002).

✛	Swastika without bent arms or even a possible Hindu-Christian syncretism of the symbol of the cross and the swastika.
दं	Dam
बो	Bo
७	Number seven
च	Ca
म	Ma
◎	In classic iconography this symbol could be associated to a shell for its inner concentric line, reminiscent of a shell's circles. In turn, the shell is a symbol attributed to Vishnu.
⊙	Unknown
⊗	Unknown
⚓	Unknown
बा	Ba

Table 2. Glossary of symbols.

Despite each group enjoying a certain amount of autonomy, a partial standardisation of style was introduced several years ago when Medge gave some useful guidelines for creating these symbols. He provides an interesting description of this change:

> As far as the code system is concerned, we did use strings in the past, not colours but cords of different colours, without applying paint with our fingers. There were seven colours and each one indicated the food's district of provenance. This system was used in Bombay when the *dabbawalas* began delivering, then the number of *dabbas* increased. For example, an area of origin was green, with a different colour for the final destination, another was yellow or white, or black [...] but it was an outdated method. The new code was my idea, when I started working. After the coloured strings, but before I arrived, there were symbols: plus, minus, dots, arrows, triangles, squares [...] This was because few *dabbawalas* can read and write, even now. But they know how to read English addresses, street names, initials, because someone else has taught them where to take *dabba*. The code is not personal, each group has a different colour. For example, my group uses green, another group has yellow, another grey; so the colour identifies the group. We abandoned triangles to adopt this coding because there are seventy-two delivery destinations. Not all *dabbawala* groups use this system, but there are two common denominators: the destination and the sorting site. All *dabbawalas* know this code identifies Churchgate, which is a big railway station. It's a large area so it's numbered from one to ten; eleven is the Marine Line. Everybody uses that numbering. If we look at this sheet we see that this is the *dabbawala* alphabet. Each letter corresponds to a group of fifteen or twenty people, and each has a letter (A, B, C, D, E). For example, in my group there may be ten people, or there are groups of fifteen or twenty, and each group has a colour. My group is green. In the group each person has a letter: he must pick up the containers from the homes and take them to the station. Each person collects thirty containers and leaves them here, at the collection point. Person B collects thirty containers and leaves them here. Person C does likewise. At the Vile Parle sorting station, persons A, B, C, etc, divide the containers according to destination: for example tiffin with the number three are all for the same destination. Then the containers depart. Another person will deliver about thirty containers all in the same area. Then there is a number for people who are given the container, and a code that indicates the building and the floor.

The *dabba* identification code was also essential in 1993, when there were bomb attacks on the railway network. On that occasion, despite strict controls during the periods following the attacks, there were no specific inspections of *dabbawalas* because, as one interviewee pointed out, in addition to the trust placed in them by the police, the *dabbawalas* themselves are the first to notice any abnormalities in their tiffin. Only the *dabbawala* and the people in his group can recognise *dabbas* that are not part of their line, thanks to the identification codes, and in such a case they would refuse them immediately. An NMTBSCT *dabbawala* explains why tiffins weren't searched during security checks:

> After the bombings there were very strict security checks but we had no problems in this respect because everyone trusts tiffin containers, so there weren't any problems. Because of the bomb attacks everyone who travelled by train was checked, but no one checked *dabbawala* tiffin containers. Everyone recognises a *dabbawala*. They start from the home, go to the station and from the station to the office. If one of our tiffin isn't right, we notice immediately. Even if we have a thousand tiffins, if one's not right... let's say someone plants a tiffin container in amongst ours, we notice immediately, because we've never seen it before. Then we know the codes found on the *dabba* but no one else can know our codes. If someone leaves a *dabba* with a bomb in it, if they do that, then we'll know. We pick up the *dabba* from home and it has our code on it. A *dabba* goes everywhere but comes back to us. Wherever it goes in Bombay, it comes back.

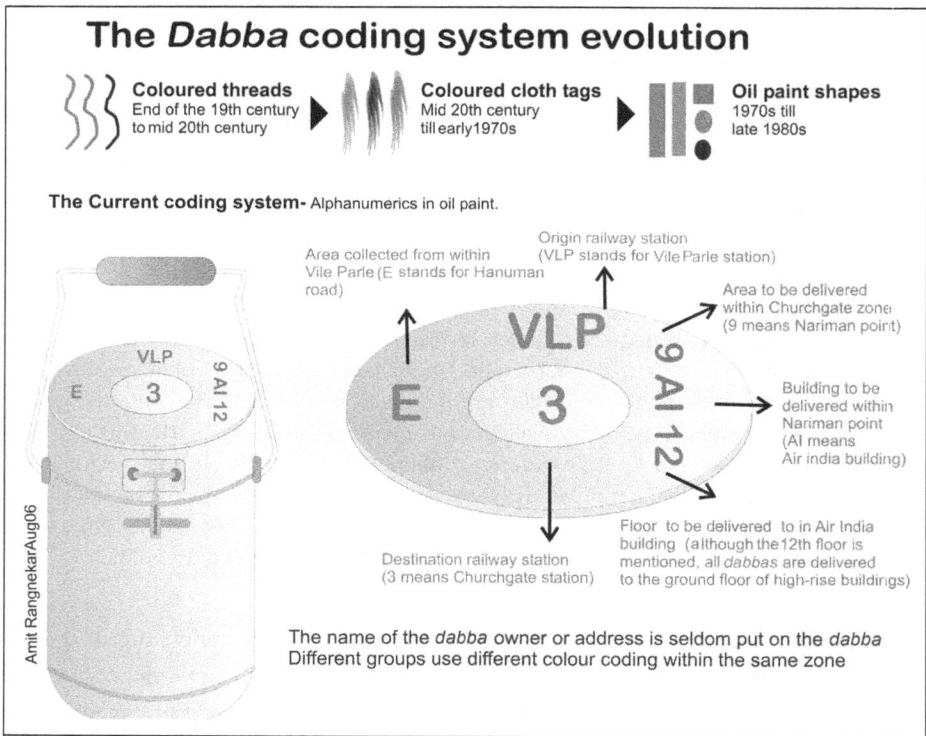

Figure 12. The *dabba* coding system evolution. Diagram by Pawan G. Agrawal, director of Mumbai Agrawal Institute of Management. By kind permission of Raghunath Medge.

Coding System

VLP   : **Vile Parle (suburb in Mumbai)**

9EX12 : **Code for Dabbawalas at Destination**

EX   : **Express Towers (building name)**

12   : **Floor no.**

E   : **Code for Dabbawala at residential station**

3   : **Code for destination Station eg. Churchgate Station (Nariman Point)**

VLP

9 EX 12

E 3

D'souza

Figure 13. The *dabba* coding current system. Diagram by Pawan G. Agrawal, director of Mumbai Agrawal Institute of Management. By kind permission of Raghunath Medge.

Apprenticeship and "practical management"

As the narrator of this short story reveals, most *dabbawalas* are illiterate or have attended only the early years of elementary schooling. Yet this does not seem to pose a problem to their tiffin pick-up and delivery work, because the basic skills required for their job rely on two main resources: possessing the physical strength needed to carry heavy tiffin baskets and being a native of the areas common to all *dabbawalas*. These two elements complement each other, because they allow workers to ground their communication in a shared language and a shared faith, as well as in a body language they recognise as their own.

Such non-verbal expressions are seldom studied in social sciences but are very important in daily personal interaction, since they allow immediate recognition among kin and kind, developing a trust that is not conditioned by verbally expressed knowledge. This "civilised gaze" is essential to social interaction, for as Paul Connerton suggests, "We always base our particular experiences on a prior context in order to ensure that they are intelligible at all".[17] The body, which basically learns the world's sensory fabric, retains a memory of our actions, incorporating individual behavioural patterns that evolve in the course of a community's collective history. This manual memory generates a different model of learning, a practical knowledge, which makes the social body movement and experience the collective expression of everyday work. Medge summarises this knowledge in his definition of practical management:

> Nowadays knowledge comes from reading books, whether they're about management, engineering or the study of social sciences. Now we just read books or a teacher explains, but that's not enough for us. Ours is 'hands-on management'. We haven't been trained. We haven't studied. There are 3,000 people in our association who still sign using their thumbs [literally: 3,000 people use a "thumbs up" to sign. This describes an informal, good-humoured gesture used to indicate the person is illiterate]. It means that they leave a thumbprint as a signature. So, to make an honest living they need a good memory. If you can read and write, and have an education, then you can write things on your computer, on your cell phone, but anyone who doesn't know how to read has to work by memorising everything. They keep everything in their heads. Like the blind who hear very well or the dumb who have better eyesight. There are about 5,000 of us and 3,000 are illiterate, but they are skilled workers. The job is to deliver food. Educated people ask questions but illiterate folk just get on with the job. The food there has to be delivered to someone, that's all they know. They don't have to worry about the rest. So there's precision in their work. If they don't work properly, they don't get paid. Without money, they go hungry. So they see only their work: with faith, devotion, strength, enthusiasm and zeal. They work with all these aspects and with stamina. With devotion, with the heart, with hard work you get something, but without prayer [because giving food is already in itself a meritorious action, a *puja*] you get nothing.

---

17   Paul Connerton, *How Societies Remember* (Cambridge: Cambridge University Press, 1989), p. 9.

An NMBTSC member explains: why we become *dabbawalas*

There are many reasons why we leave the village to come to Bombay. Some have no choice, some do, some come to work and some for pleasure. In this job the point is to have time, not brains or talent, because some of us are illiterate, some have been educated, but those who are in the service network, whether educated or illiterate, all work the same way. Once they've learned the job, once the line work is explained, even those who have no education can do it well. Anyone can be perfect for this. To do other work you need to be educated, there are limits, you need to have reached tenth or twelfth class. But here, even those with no education can earn up to 5,000 rupees. Those who live in the village are not very well educated, so they came to Bombay to earn a wage because even people who are not educated can earn well. People who live in the village are used to working from the start, both hard work and slow work: for them there's no difference. They just need the money, the money for the family. Who can manage on 1,000, 2,000, 2,500 rupees. It won't fill the belly, so there has to be another job. They like tiffin work so they move here. And if someone doesn't like tiffin work, they don't come, they go somewhere else and do another job.

Cooperative competitors

> Ten people work in my group. Each of them has bought his own line. They have the same authority as me. We can only deliver the *dabbas* if we work with each other. Give a helping hand. Like a government coalition, right? The government has policy, so do we. By uniting, we can continue to work. But governments fall. We don't, because we know that another's loss is also my loss. That's what we think and it won't change.
>
> — NMTBSCT *dabbawala*

The shared aim of each *dabbawala* team is to deliver the customer's lunch on time. The group is organised internally so that if one of the *dabbawalas* is absent, he is replaced by a colleague. Moreover, the system requires a mentality that promotes the sharing of responsibilities with intensified interpersonal cooperation. Medge describes this attitude with a metaphor: "the five fingers of a hand alone cannot do much, but if they work together they can grasp, embrace, grip, and make hundreds of gestures". *Dabbawalas* therefore manage their business without interfering with that of others, since they know that the fortune of each person is closely connected to the system's overall success. The association's structure and the type of contract applied to the *dabbawalas* also foster "cooperative competition". This means the groups compete amongst themselves to acquire more customers for their own lines, and cooperate with co-workers in the group to increase profit and, at the same time, membership for the NMTBSCT. A *dabbawala* talks about this competition:

> There's a lot of competition amongst groups, even more than there is amongst politicians. The competition is for tiffins. If someone has tiffins to be picked up, they phone for a *dabbawala*, so you have to get there first to get that tiffin. Just think, I want to pick up tiffin before the others, so I arrive and I say: 'I'm so-and-so's brother, I'll take that *dabba*'. There's so much competition. *Dabbawalas* may look a bit daft, but they aren't: they've got their heads screwed on. They'll stop at nothing to take your *dabba* and once they've got it, you can't get it back. But there's no fighting. Once the *dabba* has been collected, you know there's nothing to be gained by arguing. If there was any advantage, then we might fight, but there isn't.

Medge, however, insists that there is no rivalry:

> If someone gets your back up, how can you work? How would we manage to deliver? If I have ten customers across Bombay, I can't supply them all on my own. If I work with all the others in the luggage car, I can take other people's tiffin from the station to be delivered in my area, and they'll take mine and deliver it. Deliveries have to be regular. If you can't manage to deliver to four customers, then you don't make a profit. If you aren't part of the group, you can pick up forty tiffins from Andheri Station, but you can never deliver them to forty customers before 12.30. Lunch is at 12.30. So you have to be involved in the planning.

Another *dabbawala* adds that co-operation is key to success:

> We come from Ambegaon and Maval, the area around Ambegaon. Before coming to Bombay we didn't know one another and we got acquainted when we started the business. No one knew anyone else before that, because we're all from the same district but different villages. Usually, if I'm at the village and a friend calls me to say there's work, if I need a job I go. When we're a man short, we call someone from the village. Now we all work in the same group. Between us *dabbawalas* of the same group, we don't compete for new customers. Not like we do with other groups. At Andheri alone there are more than a hundred groups but in all Bombay there's only one association.

Competition between teams rarely turns into open conflict, but there may be problems between different groups on a train or on the platforms when they are carrying heavy baskets or bicycles. When this happens, a chain of communication is triggered from *dabbawalas* to *mukadams*, and then from the directors as far as the president, if more complex issues are to be resolved. Conflicts are never settled by recourse to external figures; instead, the association's own rules of conduct are implemented, like a "mini government" with legislation and sanctions. Medge explains what happens when problems arise:

> The *dabbawala* first calls his group leader, then the director of the reference zone is told, and he calls me in the office. If I'm not in the office, he calls my mobile phone. Then we go there. *Dabbawala* problems may be with the BMS [a municipal waste collection service] or bicycle parking. They sometimes walk across the tracks so the railway police get involved. If someone forgets their baggage pass at home, they have to find the inspector [otherwise he does not let the *dabbawala* through]. If there's any argument the police want them brought to the office [sometimes asking for money to resolve the situation]. The groups squabble amongst themselves, for competition, for money matters. One thing leads to another and there's a fight. Someone cracks a joke and someone else gets offended. If there's a dispute, a problem with the BMS or the police, or with the public, if someone on the train treads on a tiffin container, there's a row and the police turn up. Our director arrives and so does the commission, and they sort it out. It's not police business or something for the courts to deal with, we run our own regime and as such have our rules and regulations, which the *dabbawalas* themselves obey. We don't involve the Indian Government or counsellors. We have defined our own rules and regulations, and whatever happens, we solve the problem alone. Whatever problem there is at work, we find a solution. We don't call a lawyer or go to court or to the police.

Cultivating empathy with the customer

Beyond the code of ethics that underpins the delivery service—no alcohol to be consumed during working hours; wearing a *topi* cap to be recognised by customers; delivering food as if it were an act of faith; in short, all the rules summarised in the expression "work is worship"—there is also a certain empathy with the customer. Through a non-aggressive attitude, the *dabbawala* connects with what is probably the most "sentimental" part of Mumbai, the bond that exists amongst members of the family when they are apart during working hours (a wife to a husband, a mother to her children), and which is expressed in the daily act of feeding.

Dabbas' customers are predominantly men, and work six days a week outside the home, spending little time with their families. The wife usually gets up early in the morning to prepare breakfast, and after her children and husband have left for school and work, she begins to cook lunch for her spouse, waiting for a *dabbawala* to arrive and pick it up. When the *dabba* reaches the place of work, the husband opens it up and enjoys home-cooked food. The meal embodies familiar traditions, recipes and flavours. Since it is the wife who usually cleans the tiffin after it is returned, she can judge how much her husband enjoyed his meal when she sees if there are any leftovers.[18]

The long relationship of trust that enables this everyday emotional connection between the city and its "bicycle runners" has its basis in the association's internal rules. In fact, when the *mukadam* recruits, the aspiring *dabbawala* has to be of proven honesty and any violators are stigmatised by the group without having to resort to other type of justice. Medge describes the bond between the *dabbawalas* and their customers:

> People have a great deal of trust in *dabbawalas*. We've been serving some of our customers for three generations. Their grandfather ate with us, then the son and now the grandson eats with us. We don't recruit our *dabbawalas* externally, they have to have a bond with the people in our district. They have to have a permanent address to the village or in Bombay, so either a contact with a *dabbawala* or with his village. That's why customers trust us. We enter the customer's home so a *dabbawala* has to be responsible. He might find himself in a situation where he could open a wallet and steal something, plant bombs. Someone might steal or he could be a Naxalite,[19] and in that case would ruin the reputation of the *dabbawalas*. We don't want that sort of person as a *dabbawala*. You must register with our association. If one of our own steals, then we go to pick him up at the village before the police get him.

While finding new customers is one of the *mukadam*'s main tasks, a first contact is more often established through word of mouth among acquaintances and there is no particular strategy. Potential customers recognise a *dabbawala* on the street by his clothing, or they meet him at work, because his job involves delivering lunch to their co-workers. It would appear that the most important aspect for a prospective customer is that they trust a neighbour who, in turn, trusts a *dabbawala*.

---

18   Marie Percot, "Dabbawalas, Tiffin Carriers of Mumbai: Answering a Need for Specific Catering", *HAL: Sciences de l'Homme et de la Société*, 2005, available at http://halshs. archives-ouvertes.fr/docs/00/03/54/97/PDF/DABBA.pdf [accessed 10 August 2012].

19   The Naxalites take their name from the Maoist revolutionary movement that has its origins in a peasant insurrection in the late 1960s, which occurred in the rural district of Naxalbari, in West Bengal. Nowadays, "Naxalite" is synonymous in everyday language for "terrorist".

An NMTBSCT *dabbawala* describes how this works:

> Tiffin-carriers are dressed in a specific way, they wear a distinctive *topi* cap, so when people who are out on errands, out shopping, see them they call them over and say: 'Hey, you, I have a *dabba* to deliver'. Or, for example, you call for a *dabba* from an office and other people working in that office will also ask for tiffin to be brought there. If you're in the office and you eat home-cooked food while others eat purchased food, they ask: 'Who brought your food from home? How do they bring it? How much is it?' So the *dabbawala* gets contacted. Then there's also the 'title' [the advertising run by local channels: a slogan with the association's phone numbers] or sometimes flyers. That's where people get the phone number and contact us. They call the offices in Andheri, Dadar or Grant Road, and come to an agreement. They contact them and say: 'This *dabba* should be delivered to this person'. That's how we get new customers.

Medge goes into further detail:

> You see, *dabbawalas* dress in a red *jhabba* [tunic], white trousers and a *topi* cap. They have their bikes with them and the bike has a bell, so they make a noise. If there's a security guard outside your company and you need tiffin service, just tell him and he'll call the first *dabbawala* who comes out of the alley or the street, to see if he can start delivering tiffin to this building, at this number. Then lunchtime is 12.30 pm to 1 pm for everyone, in every office, factory, company, data entry desks in public offices. Then they all eat together and some of these people have tiffin delivered. So our *dabbawala* goes into the canteen. Lunch is at the same time everywhere in India, so everyone goes to the canteen together. Then if someone decides to order tiffin, they give the *dabbawala* a home address. The *dabbawala* gives it to his *mukadam*, who arranges pick-up. Anywhere you order tiffin, whether at home or at the office, my man will deliver. So tiffin is growing.

Intergenerational ties

The children of *dabbawalas* who have grown up in Mumbai often do not want to continue the work of the family. This resistance depends on several factors: often they are better educated and therefore aspire to professional prospects that were unthinkable for their parents. They also may want to withdraw from a job that appears tiring and socially degrading. The *mukadams* and the association's executive committee are not particularly concerned by this, however, because for every young Mumbaite who says no, there is another youth arriving from the old villages who wants to join the network. Indeed, it seems that these continual migrations towards the city are the lifeblood of the association's identity, characterised by rural virtues and a belief in Varkari devotional expressions. Although Medge praises these new recruits, he is aware of generational shifts: "The new boys do the *dabbawala* job well, even if previous generations worked better than the new ones. The youth don't have the stamina to work that we had. We used to carry tiffin on our heads and shoulders, now it's all done by bicycle. Now they're using technology—mobiles, trains, bicycles".

Another NMTBSCT *dabbawala* talks about the differences between the generations:

> Everyone has their own opinion about whether their children should come into the job. Now we need to see how the new generation will grow up. Someone's son will be a *dabbawala*, someone else's will be a doctor or an engineer, and someone else again won't do anything at all, while some will go back to the village, because whatever a man thinks may not be how it turns out. A person thinks ahead, but doesn't go ahead, because they will become what their *karma* has in store for them [*karma* taken to mean actions to be performed in life to achieve something or become someone]. Everyone thinks of moving up but it's not always the case. I think my son will be an important person, but who knows if this will really happen. If it's not in his Sanskriti [cultural baggage, learned from parents and family], if his *karma* is not to be, he won't become what I think or what I teach. Maybe he'll inherit the line from me, sell up and go back to the village. Could be. My father had no tiffin lines. I got the line after I arrived in Bombay. I could even just sell up and leave.

The generational turnover in the association is required to ensure the elderly return to the village and retire. During the 2007 fieldwork, Medge was considering insurance for *dabbawalas* that would provide compensation for any accidents at work and a minimum pension. As things stand, however, the livelihood of the elderly still relies on the presence of an extended family living in rural areas where the *dabbawalas* retire at the end of their working lives.[20] One *dabbawala* discusses the importance of inter-generational care:

> When people in Bombay aren't able to work any longer, they return to the village and work in the fields. It is work handed down over the generations: the son takes over from the father. There are no pensions and children take care of their parents. I don't know when. There's no fixed time. As long as I can work and there is work, then I'll carry on. That's how it is. The state gives me nothing. In our villages or at home we have our children, don't we? Maybe the son of a brother will work.

---

20   During my stay in the villages of the district of Pune, where most *dabbawalas* come from, I was a guest in Medge's home. His family—his father's two wives and an uncle, himself a retired *dabbawala*—live in a simple house surrounded by fields.

Medge says that when he retires, he will go back to his village:

> Bombay is such a costly place and it's up to the new generations to work. It takes strength to work and the elderly don't have the strength to live in Bombay. That's what the village is for. It takes three hours by train as it's about 150km from Bombay. When there are public holidays, Sundays [...] I go there to meet my family and come back to work on the Monday. There's a problem; how do you find a large apartment in Bombay? So people in the group take a *kholi* [a small room] to share. Then live together, earn the money to send back to the village. My father did that too.

Beyond technology: the railway network as a mental map

The *dabbawala* service draws its strength from a rational philosophy, an architecture of values that sustains a long-lived business based on minimum technology. These characteristics fit into the Mumbai way of life for two reasons: on the one hand, the *dabbawalas* convert their source values into knowledge of the urban territory and thus connect with the desires of people at an essential, intimate, and emotional level;[21] on the other, the world of "small things" that they represent finds the widespread transport network, in particular the railways (with their low environmental impact), to be the means for reaching every customer. One *dabbawala* emphasises the significance of the railways in his work:

> We have a lot of work on the Western Line from Virar to Churchgate. This line is very important to us, while the Thana-Dombivali-Kalyan aren't so important. If these lines don't work out there aren't many problems for our network. At most, we go up to Thane when we must, if the train's on time... But the Western Line, from Virar to Churchgate, is very important.

Every day, the urban rail and bus services are packed with commuters, who thus come into contact both physically (just taking a trip on any train will reveal what it means to be surrounded by a crowd like this, constantly jostling to get on and off) and mentally (since travelling together requires some mutual arrangement, a code of survival and reciprocal respect). This implicit code of conduct allows residents to access streets drawn on the Mumbai map, defined by stations, architecture, houses and signs that

---

21   Pandit (2007).

are all part of the sphere of social interaction.[22] The latter demonstrates how power is stratified by class (first and second class compartments), gender (men and women), or bodily expressions (it is important to avoid physical contact with other passengers). Moreover, the city's employment stratification is also clear: the packed rush-hour compartments reveal the professions of passengers. Office workers, teachers, factory workers, labourers and domestic help all usually catch trains between 7.30 and 8.30 am; sales staff going to the big shopping centres and those who do not have jobs in the public administration travel later. The railways connect the points of urban geography and are a kind of mental map for Mumbaites, allowing a linear view of the city, recognising it by the stations, the huge junctions that connect various rail lines and offering safe transit despite the endless changes in urban toponymy.[23]

The railways function as a relational nexus, an antidote to social fragmentation. Not only do they connect space and people, they also weave the threads of the many lives being played out in Mumbai into a tapestry that its residents can recognise and understand.[24] Their social role is entwined with the very nature of the city, which is neither that of a cultural "melting pot", nor of a "salad bowl" or a "mosaic"—to quote some of the metaphors commonly used by scholars to describe culturally complex human urban aggregates—but it is rather a "connective nature", experienced and reproduced on a daily basis by millions of people.

The *dabbawalas*, like most of the people of Mumbai, make this means of transport their most relied upon technology. The wooden basket is still a key device for delivery, the bicycle has only been in use for ten years and the most recent conquest, the cell phone, facilitates communications between colleagues in a line. The aim, repeatedly reaffirmed by the president, is to limit the expenses that could only be sustained by increasing the cost of the monthly subscription. One NMTBSCT *dabbawala* talks about the way that technology has changed the ways in which the *dabbawala* network can operate:

> Before, the real problem was that there weren't any bicycles, there was a wooden basket where all the tiffin was stacked. We carried the baskets

---

22   Jim Masselos, "Defining Moments/Defining Events: Commonalities of Urban Life", in *Bombay and Mumbai: The City in Transition*, ed. by Sujata Patel and Jim Masselos (New Delhi: Oxford University Press, 2003), pp. 31–52 (p. 32).

23   Ibid., p. 39.

24   Manu Goswami, *Producing India: From Colonial Economy to National Space* (Chicago: University of Chicago Press, 2004).

on our heads and we walked. But then came the bicycle, then trains and other commodities to make work easier, so that's why we like our job: it has become easier. Now, with cell phones you can find out if you've missed a *dabba*. For example, I've ten men and we're all in contact. One at Andheri, a second at Dadar, a third at Churchgate, one goes to Ghatkopar. So if I miss a Ghatkopar *dabba*, I ring my man there and he can take that *dabba*. He'll take it to Ghatkopar Station and from there another man will deliver the *dabba* at Dadar. From there another man takes the *dabba* and goes to Andheri. For instance, if someone misses the 10.30 am train, the slow train, which goes to Churchgate, he can take the fast train and we know he'll get to Vile Parle or to Dadar. So to catch up with the slow train, I go on the fast train and I catch up with him at Dadar. If I don't catch the slow train to Dadar, then I catch up at Bombay Central; if I don't catch up at Bombay Central then we catch him at Churchgate. The hour's leeway I have becomes fifteen to twenty minutes and if I can't find him I phone and say to those who have missed the train: 'Get off at Dadar'. In the meantime we have other tiffin and when we get there we hand over to the others who have the numbers. So with a mobile phone, work has improved, because we know if someone has missed a train, if someone has to get off at an earlier or later station. In a word, with a mobile it's become a bit easier. Before it was different. You didn't even know if a person had gone to work or not. By phoning you find out [...] Customers also know if the *dabbawala* is at work or not. If he's delayed he calls and we let them known he's on his way.

Not even the monsoon stops *dabbawalas* delivering, although at times high water prevents trains using their usual track. In this case, as an interviewee explained very precisely, the service stops and the *dabba* is taken back home:

When the trains were stopped for rain, we'd bring tiffin back. If the train had gone on, we'd have lost a bit of time. For instance, if a *dabba* has to arrive at 1 pm, it'll get there for 1.15 pm. But if the train stops, like when it floods with rain, then the *dabba* goes back. Any such loss is the customer's, but they don't complain because they know the *dabbawalas* have had this kind of problem. It's not intentional: if the train stops, we can't take the *dabba* as far as Churchgate. If you're close, then that's OK. But if it's from Churchgate to Grant Road, at Mumbai Central or even further than this, then we can't go by bike up to there. And even with a bike, how can we get through the streets if there's high water? So we take the food back to the home. Just like we picked it up, we take it back. If the train stops, then our work stops too.

The *dabbawalas* base the entire service on the use of suburban trains, because deliveries are limited primarily to areas where a track is installed. The railway network comprises three lines: the Western Line that passes through the western suburbs and extends from Churchgate to Virar; the Central

Main Line which passes through Thane and the central suburbs, and runs from Victoria Terminus (or Chattrapati Shivaji) to Kalyan (this is the line that meets the Western Line at the Dadar junction); and the Harbour Line from Victoria Terminus, branching at Vadala Road, with one branch going towards Panvel and the other towards Andheri. The *dabbawalas* celebrate the railways every month with a *puja* (excluding June, July and August because of the monsoons). Through this prayer, the sacralising expression of daily life, they give thanks to this valuable means of communication that allows the service to prosper by delivering lunch to customers. It is no coincidence that a popular Mumbai saying is: "If the train is the lifeline of the city, the *dabbawalas* are the food line". An NMTBSCT *dabbawala* explains the role of *puja*:

> Mr Bacche started *puja* in every station in 1952, so we too have *puja* in every station. For example, Andheri, Dadar, Marine Line and Churchgate all have *puja* every week, one week in one station, the week after in another. So in a year, every station has *pujas*, because there are so many stations. At least thirty to thirty-five stations so there are thirty to thirty-five *pujas*. We had *puja* at Andheri on 10 March, where we sang *kirtan* [a ceremony of choral singing of devotional hymns to the gods]. There was *puja* at Dadar, Khar, and Santakruz. Now there will be no further *puja* until next year because during the rainy period we don't have it. When the rains stop, *puja* starts up again. We have *puja* for *Satnarayan* [one of the names of Lord Shiva] and we also offer food [real food, not just sweets but a handful of millet]. The funds for *puja* are collected by our one hundred members: each *dabbawala* gives thirty rupees per month and each area collects money for their own district *puja*.

Conclusions: Tastes and Cultures

The sphere of taste [...] is stratified. The first level is located towards the periphery of our mental life, where we find the various qualities of taste, which may be associated to a feeling of pleasure or unhappiness. At a second level, which is much deeper and cannot be assimilated to the sphere of simple sensorial qualities, we will find our personality and the unique, specific value it glimpses in the life around it.

— Eugène Minkowski[1]

The little *dabbas* that travel daily through the crowded streets of Mumbai can be thought of as narrative devices that describe the cultures of the city.[2] They are also a gauge of the transformations at play in urban food supply and acculturation that have always characterised Mumbai—a *masala* of ingredients, flavours and commensality. As ethnographic research has taught so successfully and as Homi K. Bhabha has pointed out, "It is in the emergence of the interstices—the overlap and displacement of domains of difference—that the intersubjective and collective experiences of [...] or cultural value are negotiated".[3] The *dabba* can also be seen as "the god of small things", an expression of a world of tiny things, ordinary things,

---

1 Eugène Minkowski, *Vers une cosmologie. Fragments philosophiques* (Paris: Aubier-Montaigne, 1936), it. trans. *Verso una cosmologia* (Turin: Einaudi, 2005), pp. 184–85. English translations of this and subsequent quotes from French and Italian sources are by the author.

2 I would like to mention an interesting installation by an Indian artist Bose Krishnamachari in the exhibition *India arte oggi: l'arte contemporanea indiana fra continuità e trasformazione* (2007–2008), hosted at the Spazio Oberdan in Milan. The artwork included several *dabbas* hanging on wires to represent the many voices of Mumbai.

3 Homi K. Bhabha, *The Location of Culture* (London: Routledge, 1994), pp. 1–2.

DOI: 10.11647/OBP.0031.04

daily events that seem unimportant but in fact overflow with meaning, apparently embracing a large symbolic heritage.[4]

A humble epistemology

When my fieldwork was completed, I tried to remember the first thing I noticed when I arrived in Mumbai and immediately the memory was of the impact of different, unfamiliar smells and tastes. It is no coincidence that this impact represents a classic case of exoticism and alienation. People are so used to these sensorial stimuli, experiencing the city and its expressions daily, that they influence "social organisation, conceptions of oneself and the cosmos, control of emotions and other areas of the corporal experience", interpreting sensorial decentralisation as a fundamental law of humble epistemology.[5] It is not always easy, however, to recognise what is concealed in a taste or smell, because food flows are as extensive and complex as they are invisible. They are, however, recognisable and inscribed as secondary forms in every human body: the traces are seen in leanness or strength; the rejection of certain foods or conversely their adulation; the practices of sharing food that presume a socialisation based on the visibility of the body and its attributes; and, lastly, in the aromas of some cuisines that infiltrate and "spice" the bodies of those who consume them. Perceptions of daily habits and practices often repeat seemingly enduring stereotypes.

---

4   The reference is to the hologram principle invented by Edgar Morin, who says: "The third is the hologrammatic principle. In a physical hologram, the smallest point in the hologram image contains almost all the information of the object depicted. Not only is the part in the whole, but the whole is in the part". See Edgar Morin, *Introduction à la pensée complexe* (Paris: ESF Éditeur, 1990), it. trans. *Introduzione al pensiero complesso* (Milan: Sperling and Kupfer, 1993), p. 74.

5   Paul Stoller, *The Taste of Ethnographic Things: The Senses in Anthropology* (Philadelphia: University of Pennsylvania Press, 1992), p. 5. On sense of taste see David Le Breton, *La Saveur du Monde: une anthropologie des sens* (Paris: Editions Métailié, 2006). In recent years there have been several publications on the sense of smell and its socio-cultural implications. Annick Le Guérer states that "the same smell that marks an individual as belonging to a group and promotes cohesion, marks the individual as alien to other groups and erects a barrier between the individual and the groups". Smell then becomes the tool, justification or simply the emblem of racial, social or even moral rejection. See Annick Le Guérer, *Les pouvoirs de l'odeur* (Paris: Odile Jacob, 1998). Further references are found in Gianni De Martino, *Odori. Entrate in contatto con il quinto senso* (Melzo: Apogeo, 1997); Alessandro Gusman, *Antropologia dell'olfatto* (Rome: Laterza, 2004); and Alain Corbin, *Le miasme et la jonquille. L'odorat et l'imaginaire social aux XVIIIè et XIXè siècles* (Paris: Flammarion, 2005).

On closer analysis, representations of food, channelled into the materiality of the body, seem to reflect successfully the dilemmas of modern times and the globalisation processes that affect our everyday lives. Often these dilemmas become dichotomies: slow food versus fast food; local food versus global food; traditional versus modern; organic versus processed.[6] These polarisations express the primal fears that humans have always had when introducing different foods into their bodies. Once again the theme of cultural diversity is presented with its inherent polysemy: what is familiar is the expression of what it is—summed up in the title of Carlo Petrini's work on these issues, *Buono, pulito e giusto* [*Good, Clean and Right*] — while what is unknown, and therefore different, is dangerous and ambiguous.[7]

Food has always been potentially harmful, so recourse to tradition, and therefore to the syllogism with authenticity, attempts to be a protective measure against possible risks. Which principle, however, do we apply in inventing the protective measures that we use? Ludwig Wittgenstein answers thus: "the principle by which all hazards are reduced, because of their form, to several that are very simple that are surely visible to humans".[8] This means finding the links that connect global food practices to those of individuals. Food becomes an indicator of cultural changes because the sequence of globalisations that has occurred throughout history created local cultures and new forms of identity, which are expressed through food, language, fashion and tangible signs of continuous, reiterated acculturation processes.

Globalisation is a relatively differentiated transformation process (or set of processes) that historians call a "differential of contemporaneity".[9] Its local manifestations are not always identical, since they are expressed by mixing indigenous elements with others foreign to the territory. Food can express the diversity of forms that globalisation has assumed and continues to adopt, especially if it can be connecyed to the dynamics of large scale

---

6   Richard Wilk, *Home Cooking in the Global Village: Caribbean Food from Buccaneers to Ecotourists* (Oxford: Berg, 2006).

7   Carlo Petrini, *Buono, pulito e giusto. Principi di una nuova gastrosemantica* (Turin: Einaudi, 2005). It is not only a dichotomous polarisation: facets of food are far more indistinct.

8   Ludwig Wittgenstein, "Bemerkungen über Frazers The Golden Bough", *Synthese*, 17 (1967), 233–53.

9   See, amongst others, Zygmunt Bauman, *Globalisation: The Human Consequences* (New York: Columbia University Press, 1998); Ulrich Beck, *Was ist Globalisierung* (Berlin, Suhrkamp Verlag, 1997); and Mike Featherstone (ed.), *Global Culture* (London: Sage, 1990).

international trade.[10] Thus the forms that seek to reduce, reject or introject the impact of globalisation on food also acquire a less incomprehensible connotation. These forms fall into three main reactions:

- *resistance,* in other words the refusal that may be expressed by restricting use of products and foods arriving from abroad, or identified as symbolising a culture extraneous to one's own;
- *hybridism,* in other words a system whereby cultural diversities interweave continuously and generate new culinary traditions. A creative act that allows elements of different expressions of food to come together in a creolisation process; and
- *appropriation,* in other words a culture's capacity to absorb external influences and convert them into something that becomes part of its own history.[11]

Richard Wilk's model makes it possible to observe the gastronomic details of the diverse forms that ingredients assume in a recipe when different cultures come into contact:

- *blending*: one of the most elementary methods for developing new forms of food creolisation. Ingredients, methods, techniques, and cooking procedures are combined to obtain new combinations;
- *submersion*: a unique way of mixing foods whereby an ingredient is submerged and absorbed until it disappears into the fusion to the point that its flavour identity is totally eliminated;
- *substitution*: a technique that replaces a specific ingredient in a recipe with a new, local component. Consequently, the original dish can be simulated even when the original raw materials cannot be found for the dish;
- *wrapping and stuffing* (adapting a filling): introduction of a local ingredient into an unfamiliar dish, allowing the final dish to be identified through the "personal" flavour;
- *compression*: the decision to elect a single recipe as the icon of an entire society. The different flavours present in a civilisation are compressed and simplified. The menu of any restaurant translates the cuisine of a territory through cultural compression;

---

10   One of the best overviews of this aspect can be found in Sidney W. Mintz, *Sweetness and Power: The Place of Sugar in Modern History* (New York: Viking, 1985).
11   Wilk (2006).

- *alternation and promotion*: indicates the way in which unknown foods begin to be sampled at different times and in ways that differ from the original context. For instance, dishes usually served as main courses are presented as snacks.[12]

Consulting cookbooks will provide an understanding of the changes occurring at the gastronomic and social levels. From this point of view, cookbooks are documents, "barometers of the society that generated them", valuable indicators that can translate a territory and the life of its community at a cultural level.[13] A cookery book will describe native plants, changes in taste brought by new ingredients, treatments for illnesses, body management, food taboos, or religious requirements. The summary provided shows only some of the many ways in which different ingredients combine to create dishes in the context of a culture different from where it originally developed.[14] The preparation of food is not stable but related to social changes, themselves linked to sensorial classifications that acquire new meanings through flavour. The new flavour is rooted in collective memory and produces shared new languages and relationships, expressing transformed socialisations and business practices. The interrelationship of these spheres in turn reinforces the new flavour and expressions connected to it.

An interesting case is the expansion of Indian restaurants in Britain: the first takeaways made their appearance in the 1950s, after a large wave of immigrants arrived from the subcontinent. Indian food was associated with curry, a spice that suggested a long history of trade and relations between India and Britain.[15] Curry became synonymous with cheap Indian food and

---

12   Ibid., pp. 114–21.

13   See P. Caccia, "La cucina nei libri. Brevissima storia dei ricettari di cucina italiani dalle origini ai giorni nostri", *Eat:ing*, September 2008. Available at http://www.eat-ing.net/attach/lacucinaneilibri.pdf [accessed 18 July 2012].

14   Great flexibility and creativity in the approach to food are not a recent phenomenon. For Giuseppe Rotilio this phase established gradually in the course of human evolution and culminated 10,000 years ago with the conversion of many populations to agriculture and stock rearing. In the contemporary world there are still isolated human groups who have adapted perfectly to their ecosystem as in the arctic and subarctic tundra, Amazon rainforest, South African, and Australian deserts. Giuseppe Rotilio, "L'alimentazione degli ominidi fino alla rivoluzione agropastorale del neolitico", in *In carne e ossa. DNA, cibo e culture dell'uomo preistorico*, ed. by Gianfranco Biondi, Fabio Martini, Olga Rickards and Giuseppe Rotilio (Bari: Laterza, 2006), pp. 83–145 (p. 85).

15   It is interesting to note that the spice synonymous with Indian food worldwide is actually an English word whose origin has been attributed to several sources. Curry probably derives from a South Indian word, *kaikaari*, or its shortened version *kaari*, indicating

Indian restaurants associated with places where the working classes could eat lunch for a few pounds. The Indian restaurant was seen as a colossal entity where India's vast gastronomic repertoire was compressed into this single ingredient, curry. Until the 1970s, "going for a curry" had a special meaning, indicating where people went after the pub closed, to eat and sober up—possibly being sick in the process. Indian restaurants exploited the ignorance of British customers, who had no awareness of the difference between one type of Indian food and another. Later, a greater knowledge of food persuaded some consumers to seek more "authentic" food, where "authentic" was associated with tandoor ovens, later ousted by the wok. Then, in the 1990s, came the *balti*, a container used in Pakistan to carry bathing water and whose original meaning was actually "bucket", but over time came to mean "metal container", then pan, usually in iron or steel, for cooking *mirpuri* dishes.

In this way, the role of the Indian restaurant in British society changed, acquiring a new class of customers following fantasies of exotic, original flavours. Restaurant names were proof of this shift. In the 1960s, names like "Maharaja" or "Last Days of the Raj" were common, an expression of the recently-lost empire; later "Taj Mahal" and "The Red Fort" conjured up India's rich history prior to years of British colonisation; and then names like "Bombay Brasserie" revealed a relationship with European culture and the influence of new cosmopolitan connections. Now, the most recent phase shows a preference for names like "Soho Spice" and "Café Laziz", where it is no longer necessary to refer to "ethnic" food and there is evident familiarity with this place where Indians have now been living for decades.[16]

The multiplicity of forms assumed both by ingredients and by places where food is served shows how globalisation has not regimented sensorial categories of taste, nor has it conformed to a uniform—mostly

---

vegetables cooked in *kari*, spices mixed with coconut. Another suggestion is that the root of the word curry is *karai* or *kadhai*, indicating the wok used in Indian cuisine. Finally, since the British occupation began in Bengal, where some dishes are called *torkari*, the name may have been shortened, anglicised, and used as a synonym for Indian food.

16   See Ziauddin Sardar and Borin Van Loon, *Introducing Cultural Studies* (Cambridge: Icon Books, 1999). In Italy, Indian restaurants began to open in the 1980s and benefited immediately from the repositioning that had occurred in other European countries. The target is a middle to upper class patron and only recently have small delis started to sell cheaper Indian food for consumption on the premises or as takeaway. On "ethnic" catering see Enzo Colombo, Gianmarco Navarini and Giovanni Semi, "I contorni del cibo etnico", in *Cibo, cultura, identità*, ed. by Federico Neresini and Valentina Rettore (Rome: Carocci, 2008), pp. 78–96.

industrial—culture that imposes likes and dislikes. Instead it has adapted to the uniqueness of each place, because globalisation is a process of continuous change worldwide, feeding on the stimuli that enter the bloodstream locally. Furthermore, the ways in which people interpret the interactions between global and local also continue to change.[17] Each social group adopts its own eating style, typical of the world around it. Without underestimating changes in consumer tastes induced by advertising, mass marketing and other global stimuli—which introduce new experiences and food preferences through a slow, gradual process—food remains tied to the taste that everyone recognises as part of their cultural heritage.[18] The taste that develops over the course of a generation or a lifetime is intimately linked with childhood, family and intergenerational relations, but also changes through the different experiences everyone acquires. Taste belongs not only to impulses triggered by sensory perceptions, but also to the order of signification, the way we interpret and shape the world. Moreover, it develops as part of historical processes that change incessantly, to the point that the importance of taste seems capable of permeating the moral sphere of individuals or the tools with which it is expressed.[19]

Mumbai and its processes of food diversities

In the twentieth century various anthropologists studied nutrition, contributing to decoupling the concept of food from its basic meaning of "nourishment" as the fulfilment of a physiological need. The nature of food as a cultural construct was developed by human communities over the centuries: what people eat is the result of human history. Our species learned to use fire, to experiment with cooking techniques, to recognise poisonous foods, to develop multifaceted dishes, and to travel and export tastes and ingredients to different places. Historical anthropology highlights the importance of the so-called "plants of civilisation", which are the food foundations necessary for the development of complex cultures: wheat in Europe and the Middle East; maize in Mexico; the potato in the Andes;

---

17   Wilk (2006).
18   See Marion Nestle, *Food Politics: How The Food Industry Influences Nutrition and Health* (Berkeley: University of California Press, 2002). For a better understanding of agri-food marketing, see Antonio Foglio, *Il marketing agroalimentare. Mercato e strategie di commercializzazione* (Milan: Franco Angeli, 2007).
19   Matty Chiva, *Le doux et l'amer* (Paris: Presses Universitaires de France, 1985).

and rice in Asia.[20] The evolution of human eating practices was a lengthy historical process in which knowledge and experiences related to obtaining food (hunting, fishing, gathering, stock-rearing, agriculture, etc) and transforming what was procured, became rooted in generations of human social customs. Foods, as well as the processes that shape their edibility and usability, become the cornerstone of extensive cultural relations affected by social, economic and political dynamics.

Fernand Braudel observes that food never ceases to travel around the world and revolutionise people's lives.[21] Indeed, many of the foods we consume daily are the result of large and small migrations, historical and political changes, and infinite human curiosity.[22] This is another reason why, in the first part of this book, I emphasised the close relationship that bound Bombay food culture to the perennial flow of migrants who chose the city as a temporary or stable destination to live and work. This relationship has not changed now that the city has become Mumbai. The cohabitation of multiple ethnic groups, languages and cultures shapes the specific Mumbai migratory model. This model confers a connecting role on the figure of the migrant from the rural hinterland, who adapts to metropolitan life in such a way that they become the go-between among different lifestyles. The constant—and not always peaceful—dialectic between the city and the rural roots of most migrants promotes new social and consumption structures that appear to make Mumbai a paradigmatic global city.[23]

---

20   Alessandra Guigoni, "L'alimentazione mediterranea tra locale e globale, tra passato e presente", in *Saperi e sapori del Mediterraneo*, ed. by Radhouan Ben Amara and Alessandra Guigoni (Cagliari: AM&D, 2006), pp. 81–92.

21   Fernand Braudel, *Civilisation matérielle, économie et capitalisme, XV–XVIII siècle. Les structures du quotidian* (Paris: Armand Colin, 1979).

22   For the history of food, see Massimo Montanari and Françoise Sabban, *Storia e geografia dell'alimentazione*, 2 vols. (Turin: Utet, 2004); Christian Boudan, *Géopolitique du goût* (Paris: Presses Universitaires de France, 2004).

23   The study of cities as epicentres of global cultural and economic exchange developed in the 1980s and 1990s, as part of globalisation studies. Large urban centres have always been the subject of study, regardless of the nation states and the "global city" concept developed concomitantly with key contemporary phenomena: the end of bipolar geopolitical order; the emergence of new methods of global governance; and post-modern condition theories. See Jordi Borja and Manuel Castells, *Local and Global: The Management of Cities in the Information Age* (London: Earthscan, 1997); Manuel Castells, *The Information Age: Economy, Society and Culture. The Rise of the Network Society*, vol. 1 (Oxford: Blackwell, 1996); Saskia Sassen, *The Global City: New York, London, Tokyo* (Princeton: Princeton University Press, 1991); and Saskia Sassen, *Cities in a World Economy* (Thousand Oaks, CA: Pine Forge Press, 1994).

The city's food is also the expression of an acculturation process reiterated over time, hallmarked by its constant dynamism.[24] The clear-cut food distinctions in Mumbai's various communities reflect the changes occurring on a global scale, where migrating people bring the ingredients and recipes of their home to the city, alongside a "migration of cultural models [...] in the absence of migrant humans".[25] In other words, a migration of cognitive, behavioural and even technological models that characterise different backgrounds: rural contexts; other Indian megalopolises; diasporic places. Another tangible testimony of this complex process is found in the phenomena of food reinterpretation and revisitation pursued by Indians living outside India.

Mumbai looks like the perfect food workshop, the destination of ingredients and culinary reworking from all over the world. This is also attributable to its geographical status as a commercial port with permanent availability of typical raw materials for different diets, thanks to goods arriving from across Asia, Africa and Europe. Migratory processes have thus helped to nurture Mumbai's food prosopography, a gastronomic physiognomy that is also the result of strategies applied by the immigrant population for adaptation and economic integration. These strategies, as already seen, also delineate "parables of identity", narratives that create, enhance, or reinvent a collective identity.[26] In the first phase, migrants retain the food profile of their birthplace, preserving eating habits and seeking out places where this diet is honoured as the quintessence of a specific group identity.[27] The later stage develops over the years, with the

---

24   The "acculturation" concept developed in the USA during the difficult period preceding the 1929 global crisis. Awareness was growing of the problems of the colonised on one hand, and on the other the problems of a society that had been thriving and in continual growth. In anthropological sciences this term indicates the complex processes of cultural contact through which societies or social groups assimilate from or are imposed with elements or groups of elements by other societies. To reconstruct the wide spectrum of ethnographic and anthropological reflections on the term see Pierre Bonte and Michel Izard (eds.), *Dictionnaire d'ethnologie et anthropologie* (Paris: Presses Universitaires de France, 1991), pp. 1–2; and Alphonse Dupront, *L'Acculturazione. Per un nuovo rapporto tra ricerca storica e scienze umane* (Turin: Einaudi, 1966).

25   David Paolini, Tullio Seppilli and Alberto Sorbini, *Migrazioni e culture alimentari* (Foligno: Editoriale Umbra, 2002), p. 12.

26   Stefano Allovio, "La 'vera carne' dei pigmei: parabole identitarie e strategie alimentari in Africa centrale", paper presented at the conference *Piatto pieno, piatto vuoto, prodotti locali e appetiti globali*, Università Statale, Milan, 2 April 2008.

27   Those who migrate with the family find that meals eaten at home assume a crucial symbolic value, as do those shared with compatriots at work, or even those consumed in the small eateries run by fellow countrymen. It is also true that being able to eat food

arrival of second generations who often modify the overall migrant project and evolve it to include a wider participation in various aspects of city life, eventually also conferring a different value on food. Eating practices progressively hybridise, interacting with local ingredients and customs.

Through the continual contribution of migrations that transform and redefine Mumbai constantly, the city shows that this hybridisation process is neither linear nor diachronic. Each day it is re-lived differently, depending on the regional and caste origins of the city's migrants, and on the special values attributed to their practices. The city moves like a body, exhibiting its food needs, and in so doing draws its own cultural map, in line with the major challenges that Tullio Seppilli defines as the "planetary globalisation of the food market".[28] This complex interplay allows a glimpse into one of the possible outcomes for eating habits in global cities. The younger generations are accustomed to eating in different ways. On one hand, at home they eat the food of their family's origins, served on holidays and at get-togethers. On the other, they eat in collective contexts—at school, at work, in public spaces—where they consume "urban food", in which the combination of different traditions is enriched by the individuality of Mumbai.[29] In view of this multifaceted interaction, young Mumbaites tend to become accustomed to the incorporation of a food model that finds its truest expression in this fusion of tastes.

The relationship with the territory articulates the character that society constructs for taste.[30] Access to certain ingredients shapes our likes and

---

from one's own home enables interior control that is impossible with exterior reality. Through the food control is exercised over the body, which conversely is difficult to reproduce in the society to which one has emigrated.

28   Paolini, Seppilli and Sorbini (2002), p. 27. In Mumbai this is glaringly evident in the presence of huge global chains like McDonalds and Pizza Hut, as well as brands like Coca-Cola.

29   It is also possible to consider two different levels: on the one hand "eso-cuisine", for non-family; on the other, "endo-cuisine" for the family.

30   There are three meanings for taste: the first is linked to the sensory-perceptual dimension and sees it as one of the senses that characterise subjective perceptions (along with smell, touch, sight and hearing); the second is an aesthetic judgment that translates individual preferences for a certain object or object class; the third refers to a social dimension and defines a given preference or trend of socially defined groups. See Giorgio Grignaffini, "Estesia e discorsi sociali: per una sociosemiotica della degustazione del vino", in *Gusti e disgusti. Sociosemiotica del quotidiano*, ed. by Eric Landowski and José L. Fiorin (Turin: Testo & Immagine, 2000), pp. 214–32 (pp. 215–16). Beyond these distinctions, the three cases can be traced back to a perspective of taste belonging to the order of signification, in other words a construction that transforms and gives meaning to the world even as it evolves socially and culturally.

dislikes, developing into a system of distinctions, discriminations, class and gender.[31] Taste is probably the most visible expression of ethnocentrism. According to Sidney Mintz, the foods we eat, the criteria by which we distinguish what is edible from what is not, and what we feel when we consume them, are interconnected and contribute to our perception of ourselves in relation to others. Thus, people who eat different foods or even similar foods but in a different way, are seen as fundamentally different and, sometimes, not even human.[32] Recently this "ethnicising" dimension of food in Mumbai (and elsewhere) has been the subject of political claims rooted in far more ancient historical and social contexts.

On the subject of ethnicity

In recent years political movements have emerged in Mumbai supporting pureness of identity based on specific regional, linguistic and caste factors. These movements have triggered a "de-cosmopolitising" process that highlights the increasing value placed on ethnicity. The adoption of the term "ethnicity" in social sciences is relatively recent and, as Glazer and Moynihan pointed out in their famous mid-1970s study, it is in "one sense a term still on the move".[33] It is applicable to "situations and processes in which the cultural difference between groups is classified, organised and communicated",[34] which derives from the concept of the "ethnic group" and means "a linguistic, cultural and territorial unit of some magnitude".[35]

Historically, the concept of the "ethnic group" or "ethnicity" proved useful for meeting colonial intellectual and administrative demands. It enabled European colonialism to achieve a definition for the various subjugated communities that differed from terms like "peoples" or "nations", taken to mean subjects of a historical destiny that was long considered a European state prerogative, and who were seen to acquire "the right to self-determination" from the time of Wilsonian diplomacy.[36] The designation

---

31   Pierre Bourdieu, *La distinction. Critique sociale du jugement* (Paris: Les Éditions de Minuit, 1979).

32   Mintz (1985).

33   Nathan Glazer and Daniel P. Moynihan, "Why Ethnicity?", *Commentary*, 58 (1974), 33–39. I am quoting A. L. Epstein, *Ethos and Identity: Three Studies in Ethnicity* (London: Tavistock, 1981), p. 167.

34   U. Fabietti and F. Remotti (eds.), *Dizionario di antropologia* (Bologna: Zanichelli 1997), p. 271.

35   Bonte and Izard (1991), p. 321.

36   At the end of World War I, the US president in office, Woodrow Wilson, gave a speech

"ethnic group" in the era of European imperialism was more or less explicitly a subordinate status to communities, often meaning they were split up and restricted to specific territories.

In the 1960s, the work of Fredrik Barth, also subsequent to historical decolonisation processes, redefined the concepts of ethnicity and ethnic groups, restoring them as basic categories that allow social players to decide on a personal status and means of identification. Barth argues that ethnicity is "a category of ascription whose continuity rests on the perpetuation of boundaries and the codification constantly renewed of cultural differences between neighbouring groups".[37] The boundary is a physical and symbolic place where social interaction between groups is channelled and where the sense of identity is perceived.

Over the last twenty years, in particular for minorities and decolonised countries, the terms "ethnicity" and "ethnic group" have acquired new meanings, connected to the rediscovery of "identity roots" that colonial rule had hidden, blurred, or repressed. Ethnicist movements (in particular African-American) appropriated the terms to condemn the socio-economic injustice and exploitation that had targeted individual "ethnicities". Historical anthropologists highlighted the processes of political and economic domination of a number of social groups over others. This led several sociologists and anthropologists to identify the use (and abuse) of ethnicity as a weapon in the power struggle between variously defined groups (in which "ethnic" features considered distinctive can vary quite extensively) who are active in the promotion of their own community interests and claims to collective rights.[38] Ethnicity thereby acquires a competitive significance that alters social reality by reifying the delimitation of cultural traits such as language, religion and physiognomy, and rendering social divisions unresponsive. Behaviour with ethnic implications cannot be deemed mere rational calculation, however, because it also includes an

---

to the US Congress acknowledging the validity of the self-determination principle as a fundamental element in the new international post-war order. It was the first globally positive sanction of a collective right, a "people's" right, able to change state boundaries along lines of supposed ethno-national uniformity. In fact, these borders ended up stranding some thirty million people on the "wrong side" of the frontier and became the harbinger of conflicts that are still to be resolved. See René Gallissot, Mondher Kilani and Annamaria Rivera (eds.), *L'imbroglio etnico in quattordici parole-chiave* (Bari: Dedalo, 2001).

37   Bonte and Izard (1991), p. 322; see also Fredrik Barth (ed.), *Ethnic Groups and Boundaries* (Oslo: Oslo University Press, 1969).

38   See Glazer and Moynihan (1974); and Abner Cohen, *Custom and Politics in Urban Africa: A Study of Hausa Migrants in Yoruba Towns*, rev. ed. (London: Routledge, 2004).

important emotional component. A mono-faceted approach to the analysis of such a dense concept risks reducing its expressive potential to a mere trivialisation of the reality it seeks to interpret.

The plurality of applications of ethnicity-related terminology highlights the cultural nature of the evolution that the various players bring about, colouring it with different meanings and encompassing complex intra-psychic and social interaction processes.[39] This construction can be seen as a classification system that allows the population to be separated or grouped into specific categories, and is generally rooted in an emotional dimension.[40] In the Indian debate on the definition of ethnic groups, sociologist Gopa Sabharwal described them as:

> ... socially defined groups based on notions of shared culture which accounts for their distinctiveness. These groups are stable and have continuity over time since they perpetuate themselves. They also often possess a distinct name by which not only do the people of the group recognize themselves but are recognized by others as such. The shared cultural components could be drawn from among the following elements: region, language, religion, caste, sect, tribe, race or some of these in combination. These identities are as is evident, affiliations of birth or ascriptive identities. These cultural identities thus are those that people are born with and are not acquired or achieved. Self-awareness of identities based on these cultural criteria is an essential component in describing ethnic identities.[41]

A cornerstone of India ethnicity is the politicisation of language because, as André Béteille points out, the definition of ethnic identity on the basis of sharing a specific language produces both cultural and political effects.[42] Modern India is organised according to linguistic demarcations that often give rise to disputes amongst dominant and minority language groups, like the issue of a national standardised language as opposed to various regional or state languages within the Indian Union; the choice of which script system to use at national level; individual territorial divisions; whether or not to recognise/promote bilingualism or multilingualism in contrast with the adoption of a lingua franca like English; the characterisation of the

---

39 Epstein (1981).
40 Ibid.
41 Gopa Sabharwal, *Ethnicity and Class: Social Divisions in an Indian City* (New Delhi: Oxford University Press, 2006), p. 249.
42 André Béteille, *Society and Politics in India: Essays in a Comparative Perspective* (New Delhi: Oxford University Press, 1992).

official language of the political class; the tensions between domestic and public language; and so on.[43]

Language-based differentiation alone is not sufficient to justify the adoption of a renewed sense of "ethnic purity" as the symbolic basis for replicating certain social categories, and the question of the creation of the Maratha "caste" (explained in Chapter Two) is a prime example. To make these differentiating regimes legitimate, a Foucaultian-type governability must be introduced. In other words, a governability that can guide human conduct by regulating its expression, whose "political language drew on the rationalities of modern bio-power regulating health, reproduction, and bodies".[44]

It is in this perspective that the study of dietary practices and representations of them, even in a political context, may be useful for sounding out the less explicit dimension of this bio-power, which also feeds on symbols of identity that these practices and representations wish to embody. Ethnic groups and ethnicity must be understood as symbolic constructions, as the "product of specific historical, social and political circumstances", which indicate not a static but a ductile reality, able to bend to change according to the circumstances.[45] Through these cultural devices a collective definition of the self and the other occurs, which enables groups to acquire an internal homogeneity, highlighting the differences between them.[46]

Bombay-Mumbai, collector of cultural diversities

For Salman Rushdie, Bombay epitomises the diversities of India: it's the point of convergence for winds that blow from west to east, north to south, and vice versa.[47] For my research it became a focal point in the analysis of interpretations of cultural diversity. Mumbai, melting pot of peoples, languages, religions and cultures, is highly typical of the great cultural

43   Sabharwal (2006), p. 26.
44   Thomas Blom Hansen, *Wages of Violence: Naming and Identity in Postcolonial Bombay* (Princeton: Princeton University Press, 2001), p. 218. For more on the governability that can guide human conduct, see Michel Foucault, *Naissance de la biopolitique. Cours au collège de France, 1978–1979* (Paris: Gallimard-Seuil, 2004), it. trans., *Nascita della biopolitica. Corso al Collége de France, 1978–1979* (Milan: Feltrinelli, 2005), p. 154.
45   Ugo Fabietti, *L'identità etnica. Storia e critica di un concetto equivoco* (Rome: NIS, 1995), p. 18.
46   Ibid.
47   Sujata Patel and Alice Thoner (eds.), *Bombay: Mosaic of Modern Culture* (New Delhi: Oxford University Press, 1995), p. xxiii.

themes that pervade contemporary global thought. It actually seems like the most suitable backdrop, not dissimilar to other metropolises, for celebrating the diversity that characterises the groups that live there, as well as the particular personality the city has acquired through the coexistence of these groups. India has always had a composite and pluralist character, nurtured by a multitude of different sources: the Vedic period, a fusion of Aryan and non-Aryan populations; the rise of the Hindu religion as a mosaic of cults, gods and goddesses, and inspirations; the presence of tribal peoples; the birth of Buddhism and the Jain religion. Not to mention the Bhakti movements of spiritual renewal within Hinduism; Sikhism; the arrival of Turko-Iranian peoples and Muslims (with relative differing schools of religious thought and exegesis); English colonisation; and subsequent colonial liberation movements.

The Indian subcontinent has therefore always expressed a strong syncretic tension that combines an enormous range of regional, music, food, language and social traditions.[48] As Das points out, India is home to 4,635 different communities, most of which have their own cultural traits: dress, language, prayer, food, customs, and so on.[49] Its 325 languages rely on twenty-five different alphabets, which derive from various linguistic families: Indo-Aryan, Tibetan-Burmese, Dravidian, Austro-Asian, Andamanese, Semitic, Indo-Iranian, Sino-Tibetan, etc. Most inhabitants are bilingual or polyglot and have religious beliefs that over time have interwoven the cultural and spiritual traits of various faiths. Thus, united India has always been based on interrelations among different spoken and written cultural traditions that communities have evolved through history. It is precisely this interrelationship that constitutes the country's most fruitful legacy.

Bombay-Mumbai is the collector of this cultural, economic and legal heritage, and as such is not immune to negative reactions as seen in the recent episodes of group violence that led to bloody clashes between Muslims and Hindus in the 1990s and 2000s. This violence is not so much

---

48   Clearly this is a summary history; a more detailed reconstruction of India's history is in Michelguglielmo Torri, *Storia dell'India* (Bari: Laterza, 2000).

49   N. K. Das, "Cultural Diversity, Religious Syncretism and People of India: An Anthropological Interpretation", *Bangladesh e-Journal of Sociology*, 3:2 (2006), available at http://www.bangladeshsociology.org/BEJS%203.2%20Das.pdf [accessed 17 July 2012]. For an interesting picture of India's cultural diversities, see Gurpreet Mahajan, "Indian Exceptionalism or Indian Model: Negotiating Cultural Diversity and Minority Rights in a Democratic Nation-State", in *Multiculturalism in Asia*, ed. by Will Kymlicka and Baogang He (Oxford: Oxford University Press, 2005), pp. 288–313.

the result of a longstanding and deep-rooted hatred among different social groups (even if they do sometimes claim responsibility for actions whose symbolic roots lie in a history of abuse or enduring discrimination), but rather a reaction to the state's equality-promoting policies and external pressure resulting from global challenges. According to Arundhati Roy, there is a direct link between the current global economic structure and the birth and rise in India—as in many other poor countries—of a right-wing nationalist political class with strong racist and fascist connotations.[50] Roy's opinion is shared by sociologist Ashis Nandy in an article on the theme of anti-Islamic hatred in contemporary India: "It is the rage of Indians who have decultured themselves, seduced by the promises of modernity, and who now feel abandoned. With the demise of imperialism, Indian modernism—especially that subcategory of it which goes by the name of development—has failed to keep these promises".[51]

In this context Muslims are perceived as easy scapegoats. Claims of collective rights on grounds of purported cultural differences frequently exploit communal hatred, legitimising it as a deeply ingrained, essential feature of human nature. In fact, such claims have little or nothing to do with an alleged clash of civilisations, but are rather the offshoot of warped intra-cultural dynamics of political conflict and social strife.[52]

Rashmi Varma makes an interesting observation that, despite Saskia Sassen's argument that Mumbai is one of the cities where the geography of international finance is mapped out, the metropolis actually only penetrated the global studies panorama after its progressive "provincialisation".[53] Varma uses this term to denote a process that Arjun Appadurai calls "decosmopolitanism" and is reflected in a series of key events in Mumbai's recent political and social evolution. This includes the formation of Shiv Sena ("Shiva's Army"), a political movement based on the *bhumiputra* ("sons of the soil") concept. The Shiv Sena claims broader collective rights for the Maratha than those offered to the non-Maratha population.

---

50   See Arundhati Roy, "War Is Peace", *Outlook India*, 29 October 2001, available at http://www.outlookindia.com/article.aspx?213547 [accessed 17 July 2012]; and Arundhati Roy and David Barsamian, *The Checkbook and the Cruise Missile: Conversations with Arundhati Roy* (Cambridge, MA: South End Press, 2004).
51   Ashis Nandy, "The Twilight of Certitudes: Secularism, Hindu Nationalism and Other Masks of Deculturation", *Postcolonial Studies*, 1 (1998), 283–98.
52   In this case, Arjun Appadurai speaks of "primordialism".
53   See Sassen (1991); and Klaus Segbers (ed.), *The Making of Global City Regions: Johannesburg, Mumbai/Bombay, Sao Paulo and Shanghai* (Baltimore, MD: Johns Hopkins University Press, 2007).

The demands are accompanied by a strong Hindutva (Hindu fundamentalist and anti-Muslim) inspiration.[54] Varma also stresses the importance of exaggerated Hindu nationalism and modification of the Hindu religion in Hindutva terms as an underlying ideological strategy for this political evolution—or rather devolution. The quest for alleged ethnic purity has gradually overshadowed the city's image as the Indian capital of hope and diversity.[55] In order to catalyse this process, the neo-racist logic of absolute cultural differences—which is to say the irreducible inability to communicate among different social groups—has been used deliberately to conceal the growing and increasingly less reversible impact of economic imbalances. Historian Romila Thapar notes:

> The new Hinduism which is being currently propagated by the Sanghs, Parishads and Samajs is an attempt to restructure the indigenous religions as a monolithic uniform religion, rather paralleling some of the features of Semitic religions. This seems to be a fundamental departure from the essentials of what may be called the indigenous Hindu religions. Its form is not only alien to the earlier culture of India in many ways, but there is also a disturbing uniformity that it seeks to impose on the variety of Hindu religions.[56]

Food is precisely the link and key filter between the outside world and the body, a daily practice that places humans in systems of significance, offers the perfect setting for the explosion of feelings triggered by uniform neo-archaic food evocations, which demand ancient, authentic food as the

---

54 The term "Hindutva", coined in 1923 by Vinayak Damodar Savarkar in his pamphlet entitled *Hindutva: Who is a Hindu?*, is composed of the Persian word "Hindu" and the Sanskrit suffix "-tva". According to Savarkar it indicates the characteristics of being Hindu, of "Hindu-ness". The founding rules of Hindutva see India as the homeland of the Hindus, who are defined as "those who recognise India as a sacred homeland". In the 1980s and 1990s the Bharatiya Janata Party produced a review of the entire history of India in order to outline the characteristics of "true Hindu culture", highlighting the features of continuity with the ancient tradition and oppression by Christians and Muslims. The Sanskrit texts, especially the Vedas, were taken as the basis of "real Indian-ness", excluding all other traditions, for instance Buddhism, those of tribal peoples, and, of course, Islam and Christianity, whose presence in India is centuries old. The main purpose of this false historical reconstruction—which made use of textbooks, maps, images and videos, distributed in schools and on the streets—was to leverage the votes of the frustrated lower classes. For more detailed information, see *Ethnic and Racial Studies*, Special Issue: Hindutva Movements in the West: Resurgent Hinduism and the Politics of Diaspora, 23:3 (2000).

55 See Rashmi Varma, "Provincializing the Global City: From Bombay to Mumbai", *Social Text*, 22:4 (2004), 65–89.

56 Romila Thapar, "Syndicated Moksha?", *Seminar*, 313 (1985), 14–22.

supreme symbol of a single defined identity.[57] There is, in fact, a relationship of circular causality between food and identity. When identity is reified, and thus fossilised, in search of the paradise lost of its origins, food is also re-interpreted, scrutinising the same maps. Norbert Elias argued that there was a shift in the threshold of sensitivity, in other words the violence of the public sphere shifts to the subject's inner self and left there to implode, because—and it seems almost overkill to remember—food always includes an imaginary, spiritual, symbolic and social power that is applied according to "the principle of incorporation".[58]

This principle has two meanings: psychological and social. The psychological significance is seen in the moment of eating something and incorporating the qualities of the food. Some researchers use this key of interpretation to analyse disgust for certain foods and preference for others; in other words they start from the primordial fear of being contaminated by pathogenic organisms or the fear of acquiring the characteristics of the food eaten. This theory has two aspects: one that has a health-hygiene matrix and is based on the modern concept of illness, whereby certain foods are consumed on the basis of their nutritional characteristics; the other refers to the magical-religious thought that James Frazer classified according to two laws of sympathetic magic, one of similarity and one of contact.[59] The law of contact posits that when contact is made with a particular food, its essence is absorbed. This is particularly applicable to the Indian context when pure food, *sattvic*, is consumed with its ability to act as the channel for spiritual growth, but also to the consumption of *rajasic* or *tamasic* (for example, meat) food, which inhibits that growth. It is believed that, if certain animals are eaten, their physical or mental traits may be acquired by the consumer, so particular caution is required.

The second variation of the incorporation principle implies that when food is eaten, its *cultural connotations* are acquired: the social rules and cooking techniques connected to it and the appropriate manners required with guests while the food is consumed. In short, all the sense elements that

---

57 Jean-Pierre Poulain, *Sociologies de l'alimentation, les mangeurs et l'espace social alimentaire* (Paris: Presses Universitaires de France, 2002).
58 Norbert Elias, *Über den Prozess der Zivilisation. I. Wandlungen des Verhaltens in den Weltlichen Oberschichten des Abendlandes* (Basel: Verlag Haus zum Falken, 1939). See also Claude Fischler, *L'Homnivore* (Paris: Odile Jacob, 1990).
59 James George Frazer, *The Golden Bough: A Study in Magic and Religion* (New York: Macmillan, 1922).

make food "sociable".[60] Humans identify themselves and others through a classification system enacted for feeding themselves.[61] In a context where this sort of food diversity reigns, is it possible that this potentially implosive incorporation is guided by fear of the Other? One possible answer comes again from Fischler, who explains that to understand the contemporary eater, it is necessary to consider the *eternal eater* who, throughout history, has always had to deal with shortage of food.[62] Managing scarcity of food for centuries has programmed our bodies to respond to certain stimuli almost automatically. Although food is now available in abundance in contemporary affluent societies and, to some extent, even in most of the less affluent, the suspicion and the need arise to reject the superfluous. The anguish applies both to the excesses of modernity and to the choice of food, a choice inherent to the omnivore's condition.[63]

However, there is not only the option of refusal: responses can also be hybrid or of appropriation, as we have previously seen. Depending on the choice made, both the nature of social groupings and forms of individual actions are delineated, especially in big cities. This is because—beyond the specificities of urban life that every civilisation has developed in response to diversity of climate, religion, customs, etc—urban areas generally develop from the need to import food from outside, a necessity that changes in relation to the characteristics of the organisation of society itself.[64]

In the great metropolises, the supermarket supply chain requires standardisation of foodstuffs. This can be seen in the progressive

---

60   By "sociability" I mean the creative way in which humans put into practice acquired social and cultural resolutions, called "sociality". Sociability demands a choice in the use of forms of communication and exchange with other individuals.

61   See Fischler (1990).

62   A chapter in Paolo Sorcinelli's book has the significant title "I mali della fame e i segni del corpo" [the evils of hunger and the body's signs]. See Paolo Sorcinelli, *Gli italiani e il cibo. Dalla polenta ai cracker* (Milan: Bruno Mondadori, 1999). Much has been written on hunger, among others see Piero Camporesi, *Il paese della fame* (Milan: Garzanti, 2000), and Sharman Apt Russell, *Hunger: An Unnatural History* (New York: Perseus, 2006). In addition, in a relatively recent Indian novel, the protagonist describes the signs of hunger on the bodies of the destitute with these words: "A rich man's body is like a premium cotton pillow, white and soft and blank. 'Ours' is different. My father's spine was a knotted rope, the kind that women use in villages to pull water from wells; the clavicle curved around his neck in high relief, like a dog's collar; cuts and nicks and scars, like little whip marks in his flesh, ran down his chest and waist, reaching down below his hip bones into his buttocks. The story of a poor man's life is written on his body, in a sharp pen". See Aravind Adiga, *The White Tiger* (New York: Free Press, 2008), p. 22.

63   Elias (1939).

64   The reference is to Henri Pirenne's classic *Les villes du moyen-âge. Essai d'histoire économique et sociale* (Brussels: Lamertin, 1927).

restriction of offering to a few main brands, in constant food stocks (and thus preservation in large quantities), in the need to guarantee consequent food safety and also mass production of food, which limits genetic variety in favour of production efficiency. In Mumbai, citizens are still mostly supplied with food by independent stores, because big supermarket chains are usually located in shopping centres out in the suburbs. Unlike most first world cities, food in Mumbai is still closely tied to street sale, in small markets — a supply system that is still pre-modern to some extent.[65] Despite the diversity of food supply chains, there is tension between the globalised dimension of food flows, and therefore meaning, and the city's specific local dimension. This tension is also expressed by the social movements which — also due to the disappearance of the traditional forms of work that accompanied the birth and development of industry — now express, above all, a reactive resistance of identity type, a possible carrier of new forms of democracy, but also of outbreaks of xenophobia and religious fundamentalism.

So it seems that there is close correspondence between the expression of violence and food anguish. As Elias writes, "changes in eating behaviour are part of a broader change of human attitudes and sensibilities".[66] The incorporation of food, which is the "founder of collective identity and, similarly, of otherness", thus becomes a vehicle for reactive responses to globalisation, because to ensure the maintenance of a culturally-oriented food system, the collaboration between similars and their distinction from the Other are required.[67]

This tension between polar opposites, between "us" and "them", clearly expressed and reproduced by food consumption, is leveraged by processes of universalisation and particularisation, homogenisation and differentiation, integration and fragmentation, juxtaposition and cultural syncretism. These dichotomous positions seem to fit in well with the gastronomic order of food rules that presupposes the endorsement or prohibition of certain classes of foods. If these rules, absorbed at a subconscious level and reproduced socially as a natural order, are violated, the ensuing offence suspends the sense of community based on shared

---

65 For an overview of forms of urban food procurement, see Carolyn Steel, *Hungry City: How Food Shapes our Lives* (London: Chatto & Windus, 2008).
66 Elias (1939).
67 Fischler (1990), it. trans., *L'onnivoro. Il piacere di mangiare nella storia e nella scienza* (Milan: Mondatori, 1992), p. 52.

identities. Consequently punishment may—or will—be demanded for those who move towards the Other or who, precisely because of this offence, are perceived and stigmatised as the Other, guilty of passing the threshold of repugnance. The offending subject is then qualified as contaminated, as disgusting, and therefore is to be cast out from the sphere of humanity to which they thought they belonged.

Mumbai: relations, form, body

Mumbai's networks are an interesting way to trace back the idea of ethnicity inspired by the city's fabric of diverse citizens. In the light of Barth's supposition that a shared culture is generated by the process of maintaining boundaries, Mumbai appears potentially united—since it is crossed constantly by markers of ethnic plurality like language, clothing and food—and yet eternally divided because those boundaries simultaneously break down and pulverise the city. The two processes are dialectically opposed, but have a theoretical synthesis in Michael M. J. Fischer's definition of ethnicity as: "a part of the self that is often quite puzzling to the individual, something over which he or she is not in control. Insofar as it is a deeply rooted emotional component of identity, it is often transmitted less through cognitive language or learning (to which sociology has almost entirely restricted itself) than through processes analogous to the dreaming and transferences of psychoanalytic encounters".[68]

Perhaps this is the dimension that is reflected in "private ethnicity"— the manner in which the family structure and intimate relations initiate codes, languages, expressions and physiognomies. I would call it a sort of "cultural genetics" that "feeds the entire communication of cultural difference".[69] Social and collective articulation of this "intimate" meaning of the concept of ethnicity interprets the ethnic difference as "a complex, ongoing negotiation that seeks to authorize cultural hybridities that emerge in moments of historical transformation":[70] a process in which the private and public dimension permeate and drive each other.

---

68   George E. Marcus and Michael M. J. Fischer, *Anthropology as Cultural Critique: An Experimental Moment in the Human Sciences* (Chicago: University of Chicago Press, 1986), p. 173.

69   Matilde Callari Galli, Mauro Ceruti and Telmo Pievani, *Pensare la diversità. Per un'educazione alla complessità umana* (Rome, Meltemi, 1998), p. 179.

70   Bhabha (1994), p. 2.

The point therefore is not so much to make an inventory of the "ethnic" traits of the various components of Mumbai's metropolitan universe, but rather to direct the topic back to two important starting points for an anthropological analysis of the city: ethno-historical imagination and bio-political governance. Indeed, if the construction of ethnicity is mostly cultural, it is equally true that once imagined it becomes very significant for the groups that claim it as a defining trait of their social or even personal identity.

Bombay can be viewed from many perspectives. According to the British, the city was a symbol of progress and prosperity, the gateway to India, a home away from home, an offshoot of England. The Marathi-speaking peoples—who were the majority in Bombay but who always occupied inferior roles and places compared to the city's other minorities—had a totally different perspective. For them Bombay was in Maharashtra but not part of it:[71] it was a foreign body, a city to be admired for its architecture, for study and business, a catalyst for social change, but nonetheless deeply alien. Rooted in this were the tensions that developed in this city, which Rushdie had seen as a celebration of "hybridity, impurity, intermingling, the transformation that comes of new and unexpected combinations of human beings, cultures, ideas, politics, songs, movies".[72] Rushdie also said that Bombay "rejoices in mongrelisation, fears the absolutism of the Pure" and represents the new that penetrates the world, while in Mumbai the reality of "mixed tradition is replaced by the fantasy of purity".[73]

The change of name claimed the power of ethnicity in the wake of an ethno-historical review process. As we saw in Chapter One, the change came from a desire to make the city Maharashtra again, both symbolically and culturally. Bhabha sees "The 'right' to signify from the periphery of authorised power and privilege does not depend on the persistence of tradition; it is resourced by the power of tradition to be re-inscribed through the conditions and contradictoriness that attend upon the lives of those who are 'in the minority'".[74] The doubt arises as to whether the community of Marathi speakers, who always speak for the majority in Bombay, was actually able to perceive itself as a "minority".

71   Patel and Thoner (1995), p. 4.
72   Salman Rushdie, *Imaginary Homelands: Essays and Criticism 1981–1991* (New York: Viking, 1991), p. 394. He was speaking of *The Satanic Verses* (New York: Viking, 1988).
73   Ibid., pp. 394 and 76.
74   Bhabha (1994), p. 3.

How did it come about that a social group that had always had the power of numbers in the city was pervaded by a desire to reclaim space through "the borderline engagements of cultural difference" anchored to a specific original identity or a particular tradition inherited or represented as such?[75] By virtue of which processes was this city (which grew thanks to its ability to "embark multiplicity without endangering its own identity") able to dispose knowingly of its own cosmopolitan metropolis essence?[76]

For a better understanding of the development of these socio-political dynamics, I would like to reconstruct a shift that begins with the terrible Hindu-Muslim clashes of 1992 and 1993—in the wake of the destruction of Ayodhya's Babri Masjid—and which are considered to be the most serious post-Partition episode of ethnic conflict in Mumbai. The Hindu-Muslim conflict dates back to 1528, when the Mughal Sultan Babur ordered the construction of a mosque on the precise spot where the Hindu population believed a temple had been erected in homage to their god, Rama. In 1853 clashes erupted at the site, so much so that a few years later the British colonial authorities built a fence around it, allowing Muslims access to the building's inner courtyard and Hindus access to the outer courtyard. In 1984, the Hindu fundamentalist group Vishwa Hindu Parishad (VHP) founded a committee to liberate Rama's birthplace and build a temple in his honour. Later, Lal Krishna Advani, leader of the Bharatiya Janata Party (BJP), took over leadership of the campaign. The tone of the debate became increasingly heated until, in 1991, the BJP won the elections in the state of Uttar Pradesh, where Ayodhya is located. The following year, the mosque was rased by a crowd of 150,000 militants, triggering clashes between Hindus and Muslims across the country. Bombay, like Delhi and Hyderabad, was in the midst of this terrible violence that caused the death of an estimated 3,000 people.

The clashes recurred in March 1993 when a series of bombs placed in strategic areas of the city killed and maimed hundreds. The responsibility for the attacks was claimed by the D-Company criminal organisation, led by the Bombay mafia boss Dawood Ibrahim, and was explained as an act of retaliation for the massacre of hundreds of Bombay Muslims in the previous year. Since then, there has been repeated fighting and bloodshed across the country.[77] These actions were the result of ideologies that deliberately sought

---

75 Ibid.

76 Francesco Remotti, *Contro l'identità* (Rome: Laterza, 1996), p. 21.

77 In February 2002 a train full of Hindu pilgrims burst into flames in Godhra, Gujarat. Muslims were accused of starting the fire and thus became the target of a pogrom,

to cause rifts among the Hindu and Muslim communities, and which may be considered an expression of a "virtually worldwide genocidal impulse towards minorities".[78] They have been described by Jim Masselos as the outcome of a "polarization of attitudes among communal lines".[79]

In a process that began in the 1960s and strengthened in the 1980s, the communities residing in Bombay were categorised on the basis of religious affiliation. Hindus were considered Indian, whereas Muslims were characterised to be outsiders, non-Indians, and therefore to be excluded. The symbolic and political construction of these two opposite categories was due not only to the Shiv Sena's strategy but also to the way in which Bombay developed in the 1960s to 1980s, and the social and economic changes that affected the city during that period. The ethnic conflict revealed the fragility of tacit social norms that defined daily interactions between the city's residents.[80] The unifying energy of popular neighbourhoods-- which were a main factor in Bombay's open and cosmopolitan character-- dissolved. Violence left the city prey to the bare form of its urban body, a conglomeration of villages separated by relationships destroyed beyond repair by the blood spilled by their inhabitants.

Bombay's social networks of urban proletariat had long supported the principle of collective participation in city life; throughout the period in which industry was the principal employer of this social class, they had been an important stabilising force in the coexistence of castes, ethnicities and languages. The industrial labour crisis progressively eroded the

---

which killed some 2,000 people and led to the evacuation of 50,000 persons. In 2006, the explosion of seven bombs planted on the Mumbai urban rail network left 186 dead and seven wounded. Responsibility for the attack was claimed by Al Qaeda Jammu and Kashmir. Although there was no certainty that the different events in various parts of the country were related, violence continued until November 2008 when terrorists in Mumbai left 190 dead and 300 wounded, and a month later when police discovered two bombs—fortunately unexploded—planted on the city's railway line. Sadly, as we were translating this book into English, a new attack took place on 13 July 2011 that killed twenty-one people and injured 140. See Ramachandra Guha, *India After Gandhi: The History of the World's Largest Democracy* (London: Macmillan, 2007); Acyuta Yagnik and Suchitra Sheth, *The Shaping of Modern Gujarat: Plurality, Hindutva, and Beyond* (New Delhi: Penguin, 2005).

78   Arjun Appadurai, "New Logics of Violence", *Seminar*, 503 (2001), available at http://www. india-seminar.com/2001/503/503%20arjun%20apadurai.htm [accessed 19 July 2012].

79   Jim Masselos, *The City in Action: Bombay Struggles for Power* (New Delhi: Oxford University Press, 2007), p. 364. In current Indian political language, "communalism" is a generic term for political strategies hinged on belonging to a community or caste, and in some cases the local requirements prevail over national.

80   Ibid.

interdependence among the different segments of the weaker population, in part due to ties with class solidarity and due to the simple aspirations for social improvement that drive a migrant proletariat. These widespread solidarities were replaced by particularist affiliations and a progressive social and political ethnicisation, whose potential for disruption was revealed when an important symbolic crisis incited daily social conflict into an extended collective violence.

More than ever, the body became the mirror of ethnic affiliations and intake of culturally-oriented food acquired a symbolic and political significance.[81] Food preferences are an expression of social self-identification because: "food is a world experience; the human being is constructed through eating".[82] Nonetheless, the food chosen for nourishment is also the product of territorial policies that govern its admissibility, availability and desirability. It is no coincidence that the Shiv Sena has urged young Marathas to become street-food vendors. In this way they not only promote vegetarian Hindu food but also occupy urban land physically and symbolically, integrating business with an active street politics. Moreover, Hindu activists have repeatedly attacked *udupi* restaurants — typical of southern India but widely present in Mumbai — because they represent the growing importance of Malayalam migration. These restaurants are also publicly associated with the danger of communist infiltration, since the Kerala Communist Party has many supporters.

Yet food in Mumbai today, above all in fashionable restaurants, also offers a type of conviviality and eating that differs from Hindu traditions. It is "sociable" food that often "speaks" English, with people meeting in smart places where men and women eat dinner together, where the middle class shows off its status symbols — all of which sits uncomfortably with the rustic rural values expressed by Maratha mythology.[83] Yet even as middle-class Mumbaites appear to be gradually embracing cosmopolitanism, they nonetheless do so without challenging their need for status demarcation.

---

81 Fischler (1990).
82 Eleonora Fiorani, *Selvaggio e domestico. Tra antropologia, ecologia ed estetica* (Padua: Muzzio, 1993), p. 17.
83 Here a paradox in Indian nationalist policy seems to be established on the Maratha perception of the body. While one of Britain's key instruments of colonial consolidation process was the classification, enumeration, and discipline of the Indian body — seen as dirty, soft, brittle and effeminate — the Shiv Sena's nationalist, differentiating politics re-cast the Mumbaite's body, demanding a clean, masculine body for men who want to be in the movement. See Arjun Appadurai, *Modernity at Large: Cultural Dimensions in Globalization* (Minneapolis: University of Minnesota Press, 1996).

And for the bulk of the urban population, the effort to define boundaries and separate communities seems to have become the true *leitmotiv* of how they perceive their city's globalisation: an obsessive drive to make their food, their bodies, and the city itself adhere to a concept of purity that is truly at odds with the history of Bombay. The city is thus forced to redefine its social identity in terms of opposites, and to adopt a logic that separates and discriminates, includes and excludes. This dialectic has always been present in the city but over the past forty years has come to express an urgency of identity that also involves a politicisation of the body, its needs and its pleasures, expressed and reinforced through the adoption of distinctive eating and social practices.[84]

On the edge of cultural movements

Against this background of political unrest, food could be seen as one of the "flows" analysed by Appadurai, which enables an interpretation of the global cultural landscape.[85] Appadurai has identified five dimensions with which to describe the disjunctions between economy, culture and politics—ethnoscapes, mediascapes, technoscapes, finanscapes and ideoscapes.[86] Here it is possible to add a sixth "flow": foodscapes.[87] In this perspective, food can be detached from its "territoriality" and the

---

84   Identity claims of this nature can perhaps be read like the type of cannibalism of the Other mentioned by Francesco Remotti. Here cannibalism is a metaphor that implies opposition and accord in two societies—Maratha and non-Maratha—to the extent that one is the other's food; yet at the same time, this metaphor envisions food and related practices as a ritual that perpetuates and renews daily choices of self-representation, claiming each group's entitlement to aspiration and even cultural monopoly and elimination of what is different. See Remotti (1996), p. 77. See also W. Arens, *The Man-Eating Myth: Anthropology and Anthropophagy* (Oxford: Oxford University Press, 1980).

85   Appadurai (1996).

86   Ibid., p. 46. "Ethnoscapes" specifically refers to the increasing mobility of persons, although this does not mean that there are no stable communities (rather, these are increasingly crossed by flows of people on the move); "finanscapes" bind increasingly fluid relationships between flows of money, politics, investment, and capital; "technoscapes" refers to the rapid diffusion of technology; "ideoscapes" are concatenated images and ideas that propose a narrative of diaspora thanks to electronic information diffusion; and finally "mediascapes" examine the electronic diffusion of ideological and political images. These offer to the viewer a world and a narrative where signs and meanings of different cultural contexts are mixed, and which—thanks to the effect of imagination—encourage construction of imaginary worlds and narration of possible lives of the Other.

87   See Alessandra Guigoni (ed.), *Foodscapes. Stili, mode e culture del cibo oggi* (Monza: Polimetrica, 2004).

value attributed to it at the local level. Food expresses a bifocal vision because it is both a material flow (ingredients and raw materials used in meal preparation) and an intangible flow (feelings of identity, nostalgia, identity, claims, membership, etc evoked by food).[88] This does not necessarily lead to cultural homogenisation because, when there is a transfer of cultural material from one place to another, it is reworked in accordance with the notions of local societies, who indigenise the transferred materials and practices. Pasta is an example. It is a dish that Italians consider to be their national glory, but during its widespread global diffusion it has been radically modified to meet local tastes.[89] Food's experiential traits change over time and transformation processes do not modify only the use but also the significance, related precisely to context meanings.

Adopting the perspective of a "foodscape" enriches a political debate that is more topical than ever: trying to find solutions to the need to protect different "local cuisines". Today the aim to safeguard the authenticity of products and flavours, while being aware of inevitable taste transformations related to contemporary social changes, involves small producers, venues, and institutions at the international level, hugging borders to define new biological and cultural food maps. The *Manifesto on the Future of Food* comes in the wake of this discussion, and proposes the international promotion of food's ecological sustainability.[90] The document condemns the inability of industrial agriculture to safeguard the planet's ecosystem adequately, because it has authorised the replacement of biodiversity with monocultures whose control is completely in the hands of a few global companies. Global-local tension in the modern food panorama turns into an increasingly clear-cut political conflict. The issue hinges on the difficulties encountered by indigenous and local cultures in obtaining sufficient guarantees not only for the availability of food, but also for public health, food and nutritional quality, and the preservation of traditional forms of subsistence tied to specific cultural identities. It is interesting that the International Commission for the Future of Food and Agriculture decided to emphasise how much agrobiodiversity depends

---

88   John Berger, *Ways of Seeing* (London: Penguin, 1972). The dual focus of modernity is the ability of social players to see near and far, the local and global, simultaneously.

89   Silvano Serventi and Françoise Sabban, *La pasta. Storia e cultura di un cibo universale* (Rome: Laterza, 2000).

90   International Commission for the Future of Food and Agriculture, *Manifesto on the Future of Food* (Florence: Arsia, 2006), available at http://commissionecibo.arsia.toscana.it/UserFiles/File/Commiss%20Intern%20Futuro%20Cibo/cibo_ing.pdf [accessed 19 July 2012].

on cultural diversity, sustaining the rights of communities to preserve their identity for future generations.[91]

The contemporary food landscape therefore reveals a precise sensitivity to the development of what Watson and Caldwell call the "cultural politics of food".[92] They are a type of politics that will integrate the collective, communal and institutional sphere with the personal and private. This politico-cultural synergy, which draws inspiration from local traditions and culture to build distinctive economies characterising territorial areas, is the tangible testimony of the *dabbawala* experience. Nor is it a coincidence that the *dabbawalas* are also activists at international level for the cultural politics of sustainable food. They regularly attend international seminars on these topics, including Terra Madre, a meeting held in Turin every two years.

Food as a gift, food as a marketable commodity

Delivering food for nearly 130 years is not a "culturally neutral" act, but a cultural phenomenon that is historically and symbolically in tune with Mumbai's transformation into a global city. In addition to being an essential solution to the demand for quick meals in a metropolis whose schedule is cadenced by the working hours of the service industry, the *dabbawala* method is, implicitly, a way of ensuring the survival of the emotional and symbolic value of food prepared by loved ones, a "familiar food". In this sense "familiar food", insofar as it is able to enhance and maintain the importance of a certain type of taste that the eater does not want to give up, is one of the modern symbolic elements that contributes to the global discourse on the value of taste that assumes universal, almost cosmological, traits. Taste becomes not only a sensory experience, but a strong sense element capable of guiding choices in daily life and relationships. It can be seen as a positive result of globalisation, a process that transforms not only the world's economic structures, but also its ethical and epistemological conceptions.

---

91   The International Commission for the Future of Food and Agriculture was established in 2003 by Claudio Martini, President of Tuscany Regional Authority, and Vandana Shiva, Executive Director of the Research Foundation for Science, Technology and Ecology, Navdanya. It consists of a group of prominent activists, academics, scientists, politicians and farmers from the north and south of Italy who work to make agricultural and food systems models socially and ecologically more sustainable.

92   James L. Watson and Melissa L. Caldwell (eds.), *The Cultural Politics of Food and Eating: A Reader* (London: Blackwell, 2005).

The food delivered by *dabbawalas* can be interpreted as a "gift" to the extent that it is able to integrate the economic dimension of barter with a complex phenomenology of a symbolic, religious, aesthetic, affective and legal order. As Marcel Mauss has argued, this food can be considered a "total social fact" because it sets in motion "the totality of society and its institutions".[93] In the case of the *dabbawalas*, the value of the gift of food is expressed through the archaic conception that draws its inspiration from the recognised scriptures of classical Indian law. These laws state that:

> The thing that is given produces its rewards in this life and in the next. Here in this life, it automatically engenders for the giver the same thing as itself: it is not lost, it reproduces itself; in the next life, one finds the same thing, only it has increased. Food given is the food that in this world will return to the giver; it is food, the same food that he will find in the other world. And it is still food, the same food that he will find in the series of his reincarnations. […]. It is in the nature of food to be shared out. Not to share it with other is 'to kill its essence,' it is to destroy it both for oneself and for others.[94]

These moral and legal foundations of the Hindu tradition are part of the spiritual architecture dictated by devotion to Varkari Sampradaya. This movement explicitly considers the gift of food as the essence of egalitarianism and of serving the Other, which is considered the equivalent of serving God. The fact of being an act of devotion capable of generating karmic rewards, the gift of food (whose delivery is ultimately a good approximation) serves as an ethical constant, capable of powering a scheme that generates adaptive behaviours and allows *dabbawalas* to adapt to change in their life and work, reproducing effective collective survival strategies from one generation to the next.[95]

However, giving is not a disinterested action.[96] For the *dabbawala* the "gift-delivery (for a fee)" image constitutes a privileged symbolic resource

---

93   Marcel Mauss, *The Gift: The Form and the Reason for Exchange in Archaic Societies* (London: Routledge, 1990).

94   Ibid., pp. 72–73. Marcel Mauss refers to the *Code of Manu* and the *13th Book of Mahabharata*, the great Indian epic poem.

95   I refer to the concept of "perpetual memory", a mechanism that perpetuates a society's fundamental behaviour patterns from one era to another. Moral values, religious beliefs and social structures are components of a perpetual memory that becomes the individual capacity for remembering collectively. I think this concept is useful when placed within a dynamic analysis in which a diachronic historic dimension can be associated with a synchronic and cultural dimension. See Kirti Narayan Chaudhuri, *Asia before Europe: Economy and Civilisation of the Indian Ocean from the Rise of Islam to 1750* (Cambridge: Cambridge University Press, 1991).

96   In the preface to the English edition of Marcel Mauss, Mary Douglas explains the non-gratuity of the gift.

for preserving a mutually beneficial alliance with Mumbai. The interest of the *dabbawalas*, i.e. "the individual search after what is useful", is collateral to the actual gift principle.[97] To understand this dynamic and also the "counter-gift", it is useful to refer to Alain Caillé's third paradigm which sees the gift as a promoter of social relations.[98] If people are to live to "produce society", as Maurice Godelier argues, then the *dabbawalas* — through the relations within their group and with customers — generate an extended community that draws its strength from the continuous circle of giving, receiving and reciprocating.[99] This allows *dabbawalas* and customers to gain equally from the fulfilment of mutual needs. This implicit dialogue, which is the expression of a shared culture, overcomes "the opposition between the individual and the group, seeing people as members of a wider tangible circle" and conceives the gift not as "an economic system but as a social system of interpersonal relationships".[100]

Considering the food delivered by the *dabbawalas* to be an articulation of Mauss's gift theory does not mean "branding the gift" like old ethnographic models, which saw Other societies as more likely to give.[101] Paraphrasing Marco Aime:

> Macro models [...] risk making reality too rigid. All of us, we 'people of the Market' and they 'people who Give', are makers of actions that can

---

97   Mauss (1990), p. 96.

98   Alain Caillé, *Anthropologie du don. Le tiers paradigme* (Paris: Desclée de Brouwer, 1994), it. trans. *Il terzo paradigma. Antropologia filosofica del dono* (Turin: Bollati Boringhieri, 1998), pp. 8–11. The three paradigms referred to are: "utilitarian", which sees the person as engaged in pursuing self-interest; "holistic", which tries to explain all actions, whether collective or individual, as manifestations of the influence exerted by social totality on individuals (part of the current paradigm of social sciences such as functionalism, structuralism and culturalism); the "gift", which attempts to overcome the limits of individualism and holism, and interprets the gift as the element that allows people to create society, namely as the former of covenants. The latter paradigm, proposed and supported by the founders of MAUSS (Mouvement Anti-utilitarist dans les Sciences Sociales), including Alain Caillé and Jacques T. Godbout, proposes to assign a value to the link between goods and services in producing social reports.

99   Maurice Godelier, *L'énigme du don* (Paris: Fayard, 1996).

100 Jacques T. Godbout, *L'esprit du don* (Paris: Editions la découverte, 1992), it. trans. *Lo spirito del dono* (Turin: Bollati Boringhieri, 2002), pp. 30 and 24.

101 Marco Aime, "Introduzione. Del dono e, in particolare, dell'obbligo di ricambiare i regali", in Marcel Mauss, *Saggio sul dono. Forma e motivo dello scambio nelle società arcaiche* (Turin: Einaudi 2002), pp. i–xxviii (p. vii). See also Paolo Sibilla, *La sostanza e la forma. Introduzione all'antropologia economica* (Turin: Utet, 1996); Richard R. Wilk and Lisa Cliggett, *Economies and Cultures: Foundations of Economic Anthropology* (Boulder, CO: Westview Press, 1996); Edoardo Grendi (ed.), *L'antropologia economica* (Turin: Einaudi, 1972); and Marco Aime, *La casa di nessuno. I mercati in Africa occidentale* (Turin: Bollati Boringhieri, 2002).

be economic or convivial, without being monolithically condemned to a given datum. It is therefore possible to live in the Market, but without being subject to or exploited by it every moment of our existence. At the same time, we live in a given society and respond to specific cultural rules, but this does not mean that there are no options outside those boundaries.[102]

This food is certainly not outside the "neutral and impersonal" monetary mesh, nor is it just the inheritance of those who live outside of the market, as the *dabbawalas* themselves reflect so accurately. It is food that is part of supply and demand, capable of nourishing a class of workers who live in daily contact with Mumbai's economic and social transformations. It is food transported by entrepreneurs who make use of diverse contractual forms governed by their capacity to deliver several meals in a public context increasingly structured by the politicisation of bodies and tastes.[103]

If it is true that there is a historically new essence in the contemporary globalisation process, then the element that distinguishes it most from the preceding Fordist paradigm is perhaps the fact that the standardisation of production practices is accompanied by an increasing diversity of the symbolic worlds that interpret these practices. This diversity is ensured primarily by the close interdependence of social and cultural norms that support the forms of associative life and incorporate economic practices that give rise to specific labour organisations in specific territories.[104] Social relations, understood as expressive relationships whose intensity increases along with the growing complexity of labour division, are catalysts of this interdependence. This intensity depends greatly on the vigour of the sense systems that prevail in a given social corpus. Whether they are systems of kinship or expressions of personal ties of an affective nature, these relationships constitute "the deep psychic substance of the symbolic world of people who become subjects and not objects of economic relations and monetary economics".[105]

Culture—understood as a process that incessantly reproduces and rearranges materiality, beliefs, family systems, forms of learning, and related systems of meaning—reacts and interacts with market categories by taking advantage of shared moral and ethical codes. The specific case of the

---

102 Aime, *La casa di nessuno* (2002), p. 158.
103 Alfredo Salsano speaks of "polygamous forms of barter". See Alfredo Salsano, *Il dono nel mondo dell'utile* (Turin: Bollati Boringhieri, 2008).
104 For economic and social practices, and cultural standards, see Giulio Sapelli, *Antropologia della globalizzazione* (Milan: Bruno Mondadori, 2002).
105 Ibid., p. 115.

dabbawalas presents an economic activity embedded in a moral perspective that is both real and ideal, utilitarian and mutualistic—the empirical proof that "culture is not separable from economics, which is part of it and its rationality".[106]

Towards a unified cosmology of taste

This case-study of Mumbai's *dabbawalas* illustrates how food practices, and the social relations that sustain them, contain traces of a far broader phenomenon—the development, in the unfolding of globalisation, of a *unified cosmology of taste* that transcends borders and incorporates differences. Rules associated with eating always reflect a general notion of the universe, a cosmology. The adjective *unitary* here does not imply the gradual standardisation of taste in a global food economy, but rather points out the need for a truly global discussion on the role of nutrition in a global context.[107] In order to lend deeper meaning to the concept of "global food", some semiotic common ground is essential, and the notion of a shared cosmology is key to the project of a common social and cultural order. It also concurs with the formation and diffusion of universally accepted ethics.

Keying this "planetary discourse on food" into a discussion of an ethical nature can help safeguard its shared, open-ended approach, thus marking its distance from the imposed (and suffered) discourses that have historically shaped global taste in the previous stages of globalisation. This discussion is vital to the way local nutritional practices interact with each other and reinforce social practices and meanings that are rapidly projected on a global scale: it is, in fact, the dialectic interaction of different communities that brings about a common cosmology of taste.

The *dabbawalas'* food delivery service reflects the drive to give shape and meaning to such a shared and "permeable" set of beliefs and practices.

---

106 Jack Goody goes on to state that the concept of culture cannot be used as a "residual category, a blanket term for the non-economic aspects of social life [...]. But we need to do so [*analyse*] not in terms of a global concept of culture but of the consideration of particular socio-cultural factors seen as endogenous to the system". See Jack Goody, *Capitalism and Modernity: The Great Debate* (Cambridge: Polity Press, 2004), p. 48, my italics.

107 Examples include the many movements active at a global level, offering diverse management of agricultural resources and different food distribution. See Raj Patel, *Stuffed and Starved: The Hidden Battle for The World Food System* (New York: Melville House, 2008).

A cosmology that is at once both the true expression and the most apt description of the city's diversity, a compass to navigate the complex plots woven into the fabric of a city like Mumbai. According to Eugène Minkowski the term "cosmology" should be understood as "an aesthetic and gestural happening" in which taste acts as the life-giving vehicle of meaning. Because *to taste* means disengaging "from the matter of our daily perceptions and preoccupations [...] taking on a singular attitude towards life and the environment" and lingering in this pleasure procures a particular attitude towards the surrounding world, a primordial attitude that is "an indispensable link between the inner and the outer world".[108]

It is interesting to try to understand Mumbai's specific contribution on the evolution of this link between the individual and societal perceptions of taste. This city is one of those global urban macro-regions whose diversity seems to mimic the wider world.[109] At the same time, this diversity makes it a privileged scenario for the unfolding of India's specific way of thinking about food (see Chapter Two). This versatility, which reverberates in further layers of meaning, is also the reason for the difficulties that inevitably will be encountered in attempting to find a unified perspective for a theme as full of nuances and meanings as that of food. Thinking of an Indian food means taking into account its commercialisation, distribution and eating practices in large cities; yet it also means considering agricultural practices, livestock farming techniques and livelihoods in rural contexts. There is also the role of the ancient scriptures and the voice of the spiritual masters to be considered, not to mention global logic and the ways in which such logic is absorbed and processed by Indian nationals residing in India and abroad.

Recovering a unifying principle in all this diversity is far from simple, but it is useful to do so to avoid falling into the trap of obligatory standardisation, of convergence towards a "global culture", a definition too vague to have empirical value. The idea of a unified cosmology of taste does not end at a "global culture", but actually probes the possibility (and perhaps the need) of harmonising cultural specificities in a shared sensitivity to the protection of food integrity and eating practices. This serves to restore all the ethical, moral, spiritual and relational meanings of food, which otherwise has been reduced to its mere ingredients by a certain gastronomic trend (which is also a constituent element of the contemporary foodscape).

---

108 Minkowski (1936), pp. 187–88.
109 Segbers (2007).

Contemporary Indian thinking offers some important indications on the specific contribution that linguistics can offer for the development of a unified cosmology of taste. Indian Humanist Attipat Krishnaswami Ramanujan has argued that the ideal way to access the social and cultural context of a country is the grammatical study of its language. In many Hindu texts, grammar represents an indispensable model for thought.[110] Anthropologist Alessandro Duranti expresses similar beliefs when he says "writing is a powerful form of classification, because it recognises some distinctions and ignores others".[111] It is possible to talk of the "power of grammatical traditions" because the transcription of a language and consequently the rules underpinning its logic constitute clues on how its speakers think.[112] It is in this perspective that language—through the arbitrary taxonomies it creates—contributes to constructing culture, and that knowledge of a language allows us to understand its speaker's values.[113]

Similarly, a fine understanding of the system of values and meanings of a specific food tradition can be achieved by interpreting its "grammars of taste". Anthropologist Ravindra Khare identified three main cultural patterns in Hindu India, with three different corresponding gastrosemantics. The first, "ontological and experiential", is related to cultural assumptions constrained within an earthly and material sphere, like the classification of food, the taboos, the intrinsic qualities of foods, daily lunch models, and dietary restrictions. The second, "transactional and therapeutic", addresses how to maintain a healthy, beneficial union of body and soul (including the prevention and treatment of possible illness through diet and medicine), recognising a relationship of reciprocity between the intrinsic properties of the food and the person who consumes it. The third is the "criticism of the world", showing the limits of the preceding patterns because it addresses reality or the illusion of the world, and the role played by food in increasing spiritual knowledge, interior contemplation, and the achievement of liberation.[114]

---

110   McKim Marriot (ed.), *India through Hindu Categories* (New Delhi: Thousand Oaks; London: Sage, 1992).

111   A. Duranti, *Antropologia del linguaggio* (Rome: Meltemi, 2000), p. 116.

112   Ibid., p. 117.

113   G. R. Cardona, *La foresta di piume. Manuale di etnoscienza* (Rome: Laterza, 1985).

114   Ravindra S. Khare, "Introduction", *The Eternal Food: Gastronomic Ideas and Experiences of Hindus and Buddhists*, ed. by Ravindra S. Khare (Albany: State University of New York Press, 1992), pp. 1–26 (pp. 7–9).

Beyond individual specificities, all three "grammars of taste" give order to the world and its future, foreshadow healing and happiness, and standardise paths of self-realisation and salvation. Each of these arguments develops specific practices, expressed in the Brahman-soul principle in the first gastrosemantics, respecting the food rituals organised according to the stages of physical and social life in the second, and pursuing the fasting, renunciation and austerity of custom of the third. But this division, which is so apparent in its written form, is not so clear-cut in Hindu life today, with all three gastrosemantics incorporated into everyday life and learned during childhood and later social interaction.[115]

It is the density of meanings that this conception of taste and food contains that make nutrition one of the fields where the greatest number of political strategies are advanced today, such as demands made in the name of Hindu identity. This relationship between identity policies and nutrition is evident during the recurrent (and increasingly common) food crises brought about by rising prices in basic commodities.[116] For example, in 1998 the 5,000% increase in the price of onions became the key element in a populist political platform. The price of onions acted as an emblem of the difficulty for the poorer classes to ensure a decent diet in a globalised economy.[117] So food, the underpinning of human existence, is now a symbolic site for political activists, and has been used to justify ideologies that are also based on the violent separation of religious-ethnic groups. The rhetoric of scarcity and the struggle to satisfy the needs of "their" citizens (co-ethnic, co-religious, etc) has been co-opted for political gain. Like any type of fundamentalism, Hindutva ideology tries to replace

---

115 Ibid.
116 The FAO summit held in Rome on 3–5 June 2008, emphasised the problem of the rapid, uncontrolled increase in food prices today. Although both developing and developed countries are affected, the consequences are far more devastating in the former. The increase in price of raw materials is linked to: business cycles in key markets; climate variability; conflicts in producer countries; exchange rate fluctuations; speculation; dumping. Other factors are the increase in prices of agricultural commodities, especially rice and wheat, which began in 2007 due to the contraction of cereals exports by major producing countries to cope with growing domestic demand; increased meat consumption in China and India, which in turn triggered a major need for cereals to feed livestock; and the growing demand for ethanol as a fuel for vehicles, which raised corn prices and gradually decreased water for crops. See the website www.mascroscan.com; and Joachim von Braun, *The World Food Situation: New Driving Forces and Aquired Actions* (Washington, DC: International Food Policy Research Institute, 2007), available at http://www.ifad.org/events/lectures/ifpri/pr18.pdf [accessed 19 July 2012].
117 Vandana Shiva, *India Divided: Diversity and Democracy Under Attack* (New York: Seven Stories Press, 2005); and Patel (2008).

pluralism of ideas and practices with a focus on separation, although such a monoculture exists only in the minds of its proponents.

I believe, however, that the great wealth that Indian thinking can offer lies in its enduring attention to ecological diversity, cultural pluralism — an intellectual tendency that is not dichotomous because it addresses the unity of all things — and, finally, the intellectual flexibility that is directly proportional to the complex social relations that characterise the Indian way of life.[118] The resulting plurality of social roles has a strong impact on the nature of actual cognitive processes and allows the development of infinite cultural expressions. Nonetheless, despite existing differences, food continues to be a component of expression, as well as an integrating identity, for the many faces of today's global citizens.

---

118 I drew inspiration from the work of Rose Laub Coser, who links modern intellectual flexibility and mental operations to the complexity of social relations. In urban contexts, where the individual is in contact with greater diversity, interaction helps to reflect on the plurality of worlds in the context. See Rose Laub Coser, *In Defense of Modernity: Role Complexity and Individual Autonomy* (Stanford, CA: Stanford University Press, 1991).

Appendix: Theory and Practice for an Ethnography of Diversities

The purpose of this appendix is to highlight the concepts and guidelines that enabled my field research and the subsequent drafting of this work. First I analyse the meaning of cultural diversity, a crucial theoretical tool for understanding the dynamics of Mumbai's current situation. Second, I propose a methodological interpretation of the concept of culture that I found useful in addressing research challenges. The nature of this approach, conceived some years ago, is not only theoretical but also empirical; in my view, it is extremely useful in seeking to understand the cultural situation of the Other. Finally, I will explain the method used for my fieldwork, for gathering testimonies and digital filming—the latter being a form of documentation that allowed me to create a file of faces, expressions and non-verbal communication (for example, gestures and body language) that would not be possible just in writing.

Cultural diversity: a polysemic concept

If cultural diversity is viewed only from one perspective, then the very concept of its enlightening and scientific capacities will be misrepresented. A semantic analysis of its definition will link it back to the various ways in which culture guides human action. The heterogenesis of cultural goals—in other words the principle whereby human actions in a cultural framework may achieve different results from those originally defined—organises forms of life, social coexistence, production, exchange, and spirituality

DOI: 10.11647/OBP.0031.05

in various parts of the world, all of which appear, to use the words of Ulf Hannerz, as a "monument to human creativity".[1]

The notion of cultural diversity was initially conceptualised by nineteenth-century Romantic particularism. Stemming from a reaction to eighteenth-century Enlightenment beliefs in the universality of knowledge and behaviour, particularism regarded national cultures as distinct entities. Consequently, each culture was to be perceived in its essence or *geist*.[2] These two initial definitions fostered the development of two different anthropological schools of thought: evolutionist and culturalist. While the former sought to find a code common to all cultures, placing them on an evolutionary path where the realisation of mutual objectives varied only in the time factor, culturalists believed that each culture was unique and that existing differences should be identified, since each reference system was associated with the context in which it evolved. The two approaches, the former basically European (British) and the latter American, served different needs. Evolutionist ethnocentrism met the practical needs of colonial and imperialist rule;[3] American cultural relativism, which emerged after racial prejudice came into question, established that differences among human groups are due to their culture and evolution through history, and cannot be attributed to race.[4] The theme of cultural diversity highlights a topic that is fundamental—I would go so far as to say it is one of the discipline's foundations—for anthropology. It has always been the case that, during their work, anthropologists have had to address the importance of diversity on two planes. One brought to light differences enabling perception of the expressions that culture assumes at different times in history; the other using cultural diversity to contribute to the

1   Ulf Hannerz, *Transnational Connections: Culture, People, Places* (London: Routledge, 1996), p. 93.

2   Annamaria Rivera, "Cultura", in *L'imbroglio etnico in quattordici parole-chiave*, ed. by René Gallissot, Mondher Kilani and Annamaria Rivera (Bari: Dedalo, 2001), pp. 75–106 (pp. 83–85).

3   The reference is to evolutionary anthropology, which was established and developed mainly in Great Britain in the mid-nineteenth century. For more on the history of anthropological thought, see in particular Alan Barnard, *History and Theory in Anthropology* (Cambridge: Cambridge University Press, 2000).

4   After starting a series of ethnographic studies on North American Indian tribes, Franz Boas abandoned the principle of a single culture in favour of the idea of plural cultures influenced by multiple historical paths. Boas theorised that history does not follow a rigid pattern of evolution but is built by an infinite, overlapping series of paths. He also took pains to demonstrate the non-scientificity of the notion of a link between mental and physical traits, a notion implicit in the concept of "race". For a discussion on the two schools, see Ugo Fabietti, *Storia dell'antropologia* (Bologna: Zanichelli, 1991).

construction of differentiation theories. These concepts took two main directions. Firstly, there was a universalist approach, in which societies and cultures were classified according to a universal scale of values and in this perspective non-European cultures were initially seen (especially in the nineteenth century) as less "evolved", and needing to aim for the conditions of white civilisation. The differentialist orientation, however, considers cultural differences as natural, biological and intrinsic.[5] These two positions are not necessarily antithetical and they often interact, for instance in the recent forms of racism that manage to combine presumed respect for differences with the idea of a clear-cut separation of cultures.[6]

As a consequence, diversity refers to the condition of someone or something "perceived" as different, in other words it refers to the difference that can emerge in appearance, language, manner, conditions, ideas, opinions, tastes, etc. By extending the meaning of diversity, it is possible to embrace multiplicity and a variety of social forms to the point of attributing them with negative aspects like malice and cruelty, which may be ascribed to the person who brings conflict or friction, or does not conform. Diversity, therefore, appears to have enormous potential (it brings debate, stimulates reflection and learning, brings better self-awareness and so on), but at the same time, barely beneath the surface, it contains that age-old human fear of anything foreign, to the point of their being considered dangerous or even evil.[7]

---

5   The theories put forward were developed by Pierre-André Taguieff, *La Force du préjugé. Essai sur le racisme et ses doubles* (Paris: La Découverte; Armillaire, 1988). On the origins of racial theories, see Walter Demel, *Wie die Chinesen gelb wurden. Ein Beitrag zur Frühgeschichte der Rassentheorien* (Bamberg: Förderverein Forschungsstiftung Überseegeschichte, 1993).

6   A classic condition is described by Alessandro Dal Lago in relation to the culturalisation or ethnicisation of migrants, a practice intended to exclude them from accessing universal rights at work and in public and institutional life. The migrant's culture of origin is seen as homogeneous and monolithic, contrasting with the culture of destination, also compact and monadic. This cultural particularism seems to be incompatible with legal-political universalism, in which everyone, migrant or not, is entitled to certain rights regardless of their origin or belonging. See Alessandro Dal Lago, *Non-Persone. L'esclusione dei migranti in una società globale* (Milan: Feltrinelli, 1999), p. 171. Gerd Baumann makes reference to the same assumption when he explains how, on an expressive and political level, two apparently contradictory expressions of culture can co-exist: the "serial process" (evolutionary) and the "existentialist". See Gerd Baumann, *The Multicultural Riddle: Rethinking National, Ethnic and Religious Identities* (London: Routledge, 1999).

7   Hans Mayer analyses, through literary sources, three emblematic figures of "jarring" diversity: women, Jews and homosexuals. Mayer starts from the premise that middle-class enlightenment has failed because formal equality before the law is not consolidated by material equality in life opportunities. Enlightenment refuses to consider isolated

The aversion to diversity was also progressively rationalised into structured forms of prejudice and discrimination, like racism, a phenomenon that has taken on various forms over the centuries. As Annamaria Rivera points out, racism is definable as:

> ... the set of ideologies, statements, conducts and practices centred around the idea that certain morphological traits, biological heritage or, more strictly speaking, the genetic makeup of an individual, group or population define their psychology, behaviour, culture and personality, and that on the basis of this kind of presumed determination, hierarchies may be constructed among human groups that might justify unequal relationships, domination, exclusion, segregation and persecution.[8]

Nevertheless, biological or genetic determinism and ensuing nineteenth-century ideas of inequality amongst races cannot, alone, delimit the modern context of the racism debate. The accent today falls on cultural or ethnic differences and defines culture in a way that is very close to (or may even replace) race. This is an almost natural aspect based on unchangeable categories, a sort of datum of origin that can determine the pureness of an individual. The logic of racism amplifies the perception of cultural differences, sorting human beings into hierarchies according to these qualities, which are systematically redefined by whoever leads the hierarchy. There are no parameters defined *a priori* to radicalise the differences since, as Albert Memmi says, racism is "an accusation of variable geometry", changing its features at its own convenience.[9]

The considerations made by the anthropological world over recent decades aim to highlight the arbitrary construction of cultural dynamics. Precisely because culture is a process-based concept of numerous specificities, anthropologists suggest it should be defined in the plural—not as a single, compact, regimented "culture", but as various, dissimilar and historicised "cultures". While the "different" worlds studied by anthropologists were distant and irremediably in another place, the problem of defining cultural universes was, paradoxically,

---

subjectivity, because even diversity must be shared in a group. This group is based on apparent human regularity that does not take into account the inequality that may exist in individuality. The rock on which enlightenment shatters is as follows. Does humanity consist of men and women, physical and mental complexions of equal value or can the monsters (diversities) that are part of it enjoy the same rights as a standardised community? See Hans Mayer, *Außenseiter* (Frankfurt am Main: Suhrkamp, 1975).

8　Annamaria Rivera, "Neorazzismo", in *L'imbroglio etnico in quattordici parole-chiave*, ed. by René Gallissot, Mondher Kilani and Annamaria Rivera (Bari: Dedalo, 2001), pp. 279–309 (pp. 294–95).

9　Albert Memmi, *Il razzismo* (Genoa: Costa e Nolan, 1989), quoted in ibid., p. 296.

more difficult from a practical standpoint, but less so in an analytical sense.[10] The "Other" was seen as something distant from daily reality, defined by its remoteness, detachment and exoticism. Now the forms of hearing, seeing, feeling and representing that were once detached from us have become those of our neighbours, the passenger on the bus, the colleague, the school mate. Clifford Geertz writes that this proximity requires a readjustment of our rhetorical habits, which does not mean cultural standardisation so much as a review of the gnoseological parameters of ethnocentrism, a tendency that has been analysed in a dual perspective.[11] One is anthropological and rational, which in practice preserves the integrity of cultural reproduction processes; the other is philosophical and pragmatic, with an outlook that reinforces the sense of belonging.[12]

Conversely, if cultural diversity is to revise the parameters of knowledge of the Other, it must be acknowledged as an element of the social corpus in which we live, which is not always a huge cosmopolitan city. Moreover, at the precise moment when we shift our gaze (a difficult but inevitable action in ethnography), we can also identify cultural diversity in an intra-cultural dimension, above all in ourselves. Martha Nussbaum highlights how a clash is occurring in many contemporary nations between people who want to live with those different from themselves in a context of mutual respect, and those who seek to protect themselves within a homogeneous ethnic and religious group. In a Gandhi-inspired conclusion, Nussbaum states that ultimately the battle for democracy is fought inside each individual, between a desire to dominate and annihilate the Other, and the choice of equality and compassion as (vulnerable) foundations for cohabitation on equal footing.[13]

---

10   Clifford Geertz, "The Uses of Diversity", in *Tanner Lectures on Human Values*, ed. by Sterling M. McMurrin, vol. 7 (Salt Lake City, UT: University of Utah Press, 1986), pp. 253–75.

11   Ibid. Ethnocentrism is the tendency to consider our own group and culture as inherently superior and, consequently, to judge other cultures according to reference schemas derived from our own cultural context. Ethnocentrism considers the customs of one culture more appropriate and humanly authentic compared to customs of other groups, and reveals an unbalanced approach to evaluation and classification. It designates a universal tendency found in the many populations ("mankind", "people", etc) give themselves and in the routine practice of creating boundaries and distinctions between groups.

12   Claude Lévi-Strauss's thoughts on ethnocentrism, expressed to UNESCO representatives in 1984, are well known. He said that in order for cultures not to dissolve into one another, they must maintain a degree of resistance and ethnocentrism that allows loyalty to the values of the group to be safeguarded, enhancing internal cohesion. The speech followed an earlier work on the topic entitled *Race and History*, which UNESCO requested from the anthropologist in the 1950s.

13   Martha C. Nussbaum, *The Clash Within: Democracy, Religious Violence, and India's Future* (Cambridge, MA: Harvard University Press, 2007).

Cultural diversity and political issues

Cultural diversity has been addressed here with reference to including otherness in the institutional and organisational sphere. From this perspective, the concept expresses its polysemic aspect to the full since it incorporates different meanings, depending on which players make use of it. Themes include identity and gender, the role of ethnic and religious minorities in society, the citizenship of migrants, the rights of second generation immigrants, etc. Despite having an array of meanings, the term "diversity" evokes the differences in cultural practices, preferences and values typical of groups that "cohabit the same space".[14] It also now subsumes the political-regulatory practices that aim to secure recognition and respect of these differences.[15] The emphasis on cohabiting the same space underscores the importance that reflections on diversity attribute to place, a physical space, and also virtual reality (the symbolic space promoted by modern information and communication technologies, of which the Internet is perhaps the strongest avatar), which seems to swallow up the mental scenarios of social interaction. Place is thus the material expression of the construction of social and cultural approximations. It is the keystone of individual and collective identity, a foundation on which to erect a sense of belonging and identification.[16] It is no coincidence that the theme of diversity is extensively applied in so-called "multiculturalism", a

---

14   Enzo Colombo, *Le società multiculturali* (Rome: Carocci, 2002). For a more recent reflection, see Enzo Colombo and Giovanni Semi (eds.), *Multiculturalismo quotidiano. La pratica della differenza* (Milan: Franco Angeli, 2007). The term "diversity" found wide application in US business strategies. Originally "diversity management" was designed to ensure equal employment opportunities for women and the disabled; now corporate law also provides for the protection of ethnic minorities. See Luca Visconti, "Diversity management e lavoro straniero: vantaggio competitivo o cerimonia?", in *Diversity management e società multiculturale. Teorie e prassi*, ed. by Luigi Mauri and Luca Visconti (Milan: Franco Angeli, 2004), pp. 11–30 (p. 16).

15   See, for example, Ida Castiglioni, *Dal multiculturalismo al diversity management. Una ricerca empirica sulla definizione e sulla misura della competenza interculturale nei servizi sanitari e sociali di Milano e della sua provincia* (Milan: Provincia di Milano, 2008), available at http://www.provincia.milano.it/export/sites/default/affari_sociali/Allegati/multiculturalismo_diversity.pdf [accessed 30 July 2012]. The text emphasizes how "diversity is an organisational application of various inclusion policies, based on a broad concept of culture that embraces diversity of gender, nationality, local belonging, physical ability, generation, and role".

16   On the relationship between space and culture, see Akhil Gupta and James Ferguson "Beyond 'Culture': Space, Identity, and the Politics of Difference", *Cultural Anthropology*, 7:1 (1992), 6–32.

political strategy for managing inter-ethnic relations that aims to achieve reinforcement and respect of cultural differences.

Multiculturalism was defined in the mid-1980s and gradually established itself as a new ideology, first in the United States and then in Europe. Its first elaboration dates back to the advent of the so-called "question of difference", a difference that would be able to produce a new social identity. It was actually in the 1960s that western society began to demand a greater sensitivity and recognition of differences in gender and age, to act as catalysts for the appearance of new collective subjects. This protest soon joined forces with accusations of imperialist leanings perceived in the dominant—western—culture. There was a progressive bonding with the ethnic and nationalistic demands inherent to the historical decolonisation processes, although often under the apparently universalist aegis of Marxist-Leninist ideology. The consequences of decolonisation included ever-increasing migratory flows that shifted active populations from former colonies to their previous metropolitan territories. Countries once characterised chiefly by emigration (as were virtually all European imperialist states) became destinations for immigrants of extremely varied ethnic and national origins. These societies thus showed an increasing resemblance to the ancient settlement colonies that later became "multi-ethnic societies".

The major cities in these countries became the heart of a debate on the coexistence of cultures, which in Europe—at least initially — rooted itself in internationalist and third-world rhetoric. In the US, Canada and Australia, however, it was inspired by a critical revaluation of the countries' past as colonies that developed from the exploitation or extermination of other peoples or minorities. India was perhaps the only part of what was then called the "Third World" to develop its own ideology and multiculturalist idiom quickly. Indeed, to some extent it was ahead of the times. The emerging "global cities" were the inevitable flipside of these collective representations, precisely because cohabitation, affection and alienation intensify in confined spaces, taking for granted a redefinition of role and belonging, not to mention the development of non-violent coexistence practices between strangers and different types of persons. The spread of new terminology indicated willingness and commitment to providing responses to the changes under way.

Multicultural ideology developed in the US following a crisis in the American assimilationist model, whose most powerful metaphor is probably that of the "melting pot", a crucible, symbolising the hope for a fusion of different cultures, with the scope of generating a new culture to take their place.[17] Although the term itself was not in common use until the twentieth century, in as early as 1782 the French immigrant J. Hector St John de Crèvecoeur conveyed the idea that the future of America would lie in creating a new civilisation, the expression of a fusion of all races.[18] The "melting pot" concept, however, also had a subtly political dimension since it proposed an assimilationist-type acculturation process. The idea was that all differences could be traced back to the specific cultural legacy of the place of origin of the immigrants, which would have to be forsaken in the name of cultural standardisation epitomised by the image of the white, Christian, English-speaking American citizen.

After World War II, the widening gap between the standard of living and the actual enjoyment of civil and social rights in the various ethnic and "racial" communities resident in the US (in particular between the African-Americans and the others, especially the so-called White Anglo-Saxon Protestants) triggered clashes that came increasingly to the public eye. The attention of the world's media to these conflicts was accentuated by the new response awakening in the country's progressive elite, inspired by the emerging political awareness of the decolonisation process in non-whites. This awareness led to the development of the civil rights movement, whose heroes and representatives became outstanding personalities in the African-American communities, but there was also strong participation by liberal whites, and in particular the youth who later formed the American New Left.

Alongside these claims based on the ideology of equality, promoted by the Declaration of Independence and the American Constitution, an

---

17   The term "melting pot" originates from Israel Zangwill's play, *The Melting Pot*, written in 1909 and staged with great success in several major American cities. Its protagonist David often uses the expression "melting pot" in the play, for instance in defining America as "God's crucible, the great melting pot where all the races of Europe are melting and re-forming". See Israel Zangwill, *The Melting Pot* (Charleston: Biblio Bazaar, 2008).

18   In the third letter of his pamphlet, *Letters From an American Farmer*, de Crèvecoeur actually poses the problem of American identity and provides his own reply: "What then is the American, this new man? [...] He has become an American by being received in the broad lap of our great Alma Mater. *Here individuals of all races are melted into a new race of man*, whose labours and posterity will one day cause great changes in the world" (my italics). J. Hector St John de Crèvecoeur, *Letters From an American Farmer* (New York: Fox, Duffield, 1904), available at http://xroads.virginia.edu/~HYPER/crev/home.html [accessed 30 July 2012].

approach developed whereby the crucial issue was deemed to be that of reaching the "perfect union" advocated by the founding fathers of the American republic. More radical movements arose, however, criticising this universalist perspective and demanding that differences and peculiarities be appreciated, putting forward the recognition of collective—"community"—rights as the reference category. Minority groups, especially from the 1970s onwards (the most obvious US references here are the black Muslims, Malcolm X, as well as more radical feminists) became increasingly overt in demanding formal recognition of specific identity differences: women, youth, African-Americans, homosexuals, etc.[19] They refused to adapt to predetermined social models built around sanctioning of specific superiorities that were intended to reproduce structures to oppress whites, the elderly, men, heterosexuals, etc.

Multiculturalist reformation thinking in the 1980s and 1990s condensed some of these examples—initially suggested in a far more radical and violent key—by using first the mosaic metaphor (the vision embraced by African-American radicals who wanted a rigid separation of communities to ensure the collective identity of each group was respected), then that of the salad bowl.[20] The latter was a more inclusive perspective that put a significant distance between itself and the melting pot ideal, stating that in the same way a salad is made up of many different ingredients, the US comprised different cultures, so while they might be mixed together in a single context (the salad bowl), none lost their individuality or their flavour. This principle stated that all minorities present on the territory could claim collective rights, condemn discriminatory stereotypes and demand a constant monitoring of how the various reference communities were depicted in the media. Dedicated strategies were developed for overcoming

---

19   Each of these groups has also developed its own approach to cultural diversity. On the issue of male-female differences, I should mention the work of anthropologist Françoise Héritier, who begins with the premise that social gender distinction is a constant factor in human history and tries to understand its invisible, symbolic roots. According to Héritier, gender categories and representations of the sexual person are cultural constructions, but they all start from a given biological reproduction distinction. While the physiological description may be required for understanding diversity, this does not mean that this data is translated in a unique, universal manner. Indeed, anthropological practice has shown how this translation varies. Héritier's definition of the oriented, although not hierarchical, conceptual relationship between masculine and feminine is "gender value differential". See Françoise Héritier, *Masculin-Féminin. La pensée de la différence* (Paris: Éditions Odile Jacob, 1996).

20   See Nathan Glazer, *We Are All Multiculturalists Now* (Cambridge, MA: Harvard University Press, 1998).

inequalities deriving from discrimination of identity, for instance so-called "positive discrimination" or "affirmative action".[21]

The pressure to reinforce collective minority identities posed major new issues, which had an impact on the political choices that each state had to implement. Pressure coming from the fringes of society required rethinking of processes for the interpretation and designation of the sense of collective life. In particular, Enlightenment-inspired universalism, based on the unifying principles of equality and fraternity, collided increasingly with the adjustment of particularism of identity for minority subjects. A crisis also arose in the notion of citizenship, or rather the equality of citizens in their relationship with the state. Above all, at the end of the Cold War, in a world that suddenly appeared smaller and more connected, but also more multipolar and interdependent, a widespread need emerged to safeguard local identities. This can be defined as a turning point in western gnoseological ethnocentrism and was itself part of an epistemological turning point that penetrated social sciences,[22] and showed how values, knowledge and truth are fundamentally relative since they depend on the cultural context that generates them.[23]

In this way, the ability to define the universal fundamentals that can represent an active reference for ethical and political decisions is called into question. Here we see one of the greatest threats attributed to ideologies built on cultural diversity, as they become political management tools for inter-ethnic relations. The defence of diversity at all costs turns into complicated disunion, cultural boundaries become distinct, and a forced promotion of each social group's unique traits seeks to establish access to rights and protections based on cultural distinctions within the community. Paradoxically, these policies reduce diversity by stereotyping the

---

21   See Mondher Kilani, "L'ideologia dell'esclusione. Note su alcuni concetti-chiave", in *L'imbroglio etnico in quattordici parole-chiave*, ed. by René Gallissot, Mondher Kilani and Annamaria Rivera (Bari: Dedalo, 2001), pp. 9–36 (pp. 12–20).

22   The turning point was preannounced in the early 1900s by Husserl, in the field of philosophy, by Weber and Durkheim in sociology, by Boas and Kroeber in anthropology, and Saussure and Sapir in linguistics.

23   I refer in particular to Geertz's interpretive anthropology. Geertz argues the need to treat cultural phenomena as systems of meaning and symbols which must be interpreted in a dual perspective: on one hand, the local context in which they are produced, on the other in relation to the researcher's mental conditioning. Interpretive studies of culture seek to identify the diversity of ways in which humans shape their existence. This epistemological breakthrough towards the meaning plays off two subjectivities in ethnographic research and processing of reality: the investigated and the investigator. See Clifford Geertz, *Interpretation of Cultures* (New York: Basic Books, 1973).

community and are often promoted by the social players themselves and the media, who create particularisms based on examples that are neither non partisan nor neutral.[24] It is expedient for social players to identify themselves in a concept of culture that brings moments of totality, unity and integrity—in Wim van Binsbergen's words, "to turn subjectively the fragmentation, disintegration and performativity of modern experience into unity, coherence and authenticity".[25] Culture becomes a promotional commodity that is able to sell differences because it takes for granted that an absolute difference exists and consequently inaugurates descriptions like "the clash of civilisations", "the clash of cultures", "ancestral tribalism" and "the culturally congenital poverty of peoples".

The unease caused by economic inequity, disproportionate access to raw materials and essentials like food and water compounds the spectre of allegedly invincible cultural diversity between one group and another. In this way, cultural diversity, contrary to the status it is afforded in anthropology, would paradoxically legitimise the reduction of complex forms of social coexistence into institutionalised structures of separation and division.[26] This tension results in endless contradictions and an insidious antagonism that often brings on violence. In times of recession and destabilisation of daily routine, this irrational attitude to cultural diversity immediately finds fertile ground for growth since it puts the idea of diversity back into "reassuring patterns of our classification system", restoring to diversity its meaning of "not conforming to what is familiar".[27] Instinct does not allow dealing with others by recognition, dialogue or curiosity, but rather by suspicion, anxiety and ostracism. The attitude in the community does not

---

24 Ugo Fabietti, "Diversità delle culture e disagio della contemporaneità", *I Quaderni del CREAM*, 9 (2009), available at http://www.unimib.it/upload/aa_fabietti.pdf [accessed 30 July 2012].

25 Wim M. J. van Binsbergen, "Cultures Do Not Exist: Exploding Self-Evidences in The Investigation of Interculturality", *Quest: An African Journal of Philosophy*, 13 (1999), 37–144, available at https://openaccess.leidenuniv.nl/bitstream/handle/1887/9525/ASC-1239806–260.pdf?sequence=1 [accessed 30 July 2012]. The article is based on the inaugural speech delivered by van Binsbergen when he accepted the Chair of Intercultural Philosophy at the Faculty of Philosophy, Rotterdam Erasmus University, January 1999. See also Anthony Giddens, *The Consequences of Modernity* (Stanford, CA: Stanford University Press, 1990).

26 See Marco Aime, *Eccessi di culture* (Turin: Einaudi, 2004). On the relationship between economic inequality and cultural diversity Walter Benn Michaels's work is illuminating, with its affirmation that cultural diversity and related identity issues are used to mask economic and class rifts in American society. See Walter Benn Michaels, *The Trouble with Diversity: How We Learned to Love Identity and Ignore Inequality* (New York: Metropolitan, 2006).

27 Fabietti (2009).

refer only to the notion of a western world hopelessly compromised by guilt for its colonial conquests, but also to a hypostasised idea of minorities, colonised subjects and victims of exploitation who are denied any chance of power except through violent rebellion. These overlapping attitudes are like a hall of mirrors, in which the highly emotive, biased images reflect repeatedly, driving use and abuse of cultural differences for political ends.[28]

A methodological interpretation of culture

In the text I used the word "culture" several times. Despite the exasperatingly self-evident and obvious way the word is used in public discourse, which Marc Augé sees as an ethnological datum in itself, it is useful to highlight that culture is a profoundly complex abstract construction that is essentially the outcome of an invention.[29] Culture creates collective and patterns that are reproduced through the linguistic forms of signification. These forms are not a representation of culture, but a tool for understanding the rationale of its construction, because for every cultural trait that defines a linguistic form, there are others that extend their reach beyond the confines marked by the representation itself.[30] In the case of my ethnographic material, the data refers to cultural meanings that take into account the whole debate surrounding this tricky concept. Some anthropologists believe that the use of culture as a heuristic category raises more questions than answers, and

---

28   This is culturalism, namely the propensity to define humans not as makers of culture but as products of it, and to describe conflicts related to the economic, legislative and social sphere on the basis of cultural difference. Charles Taylor's theory is central to criticism aimed at multiculturalism as a perspective that encourages this exaltation of cultural differences. Taylor believed that the principle of coexistence could not be mutual exclusion and fragmentation of social groups; on the contrary, there should be the presence of equal democratic institutions allowing citizens complete realisation under the umbrella of universal rights that safeguard cultural individuality. See Amy Gutmann (ed.), *Multiculturalism: Examining the Politics of Recognition* (Princeton, NJ: Princeton University Press, 1994).

29   Marc Augé is discussed by Denys Cuche, *La notion de culture dans les sciences sociales* (Paris: Éditions La Découverte, 2001). See also Francesco Remotti, *Cultura. Dalla complessità all'impoverimento* (Rome: Laterza, 2011). The invention of culture is concomitant with the invention of the Other; see Mondher Kilani, *L'Invention de l'autre. Essais sur le discours anthropologique* (Paris: Payot et Rivages, 1994). On the invention of culture, see Roy Wagner, *The Invention of Culture* (Chicago, IL: University of Chicago Press, 1981). On the construction of culture, see van Binsbergen (1999).

30   I draw inspiration from the exhibition *L'immagine emerge dall'intersezione delle pieghe* [the image emerges where the folds intersect] by the artist Barbara De Ponti, who exhibited her work in a group show called "Simmetria personale" at Bollate's Fabbrica Borroni, 2008.

therefore should be set aside. I believe that in so doing, we would fail to allow the ambiguity and ingenuousness concealed in the development of cultural anthropology to come to light.[31]

Culture in the anthropological sense has no classifying or descriptive intent: it stands as the logic that allows the essence and basis of distinctions to be understood.[32] There are three main characteristics in culture: it is *learned, shared* and has a *highly symbolic nature*.[33] In the course of their existence, humans learn the wide range of expressions that actually make them human. Throughout their lives they learn to speak, to move and to think of other human beings.[34] This learning process requires strong cultural flexibility: in order to stay alive itself as a system of signification and decoding of the complexity of human experience, it must renew from one generation to the next. Jack Goody argues that:

> human societies consist of interlocking chains of generations that both transmit and innovate and human cultures consist of chains of interlocking communications; innovation would be impossible if language remained substantially the same over time, enabling inter- and intra-generational communications to take place. And communication necessarily involves a level of understanding the other, so that the new is almost always, in a sense, the transformation of the old, carrying along the 'traces'.[35]

---

31   Cuche (2001).

32   Francesco Remotti, "Cultura", in *Enciclopedia delle Scienze Sociali*, vol. 2 (Rome: Istituto Enciclopedico Italiano, 1992), pp. 641–60.

33   I consulted Piero Vereni's very useful website and I am grateful to him for making his lectures and teaching material available to everyone. See http://pierovereni.blogspot.com [accessed 30 July 2012].

34   The body translates the different eras and societies, and can therefore be regarded as a kind of cultural filter. We can say that the body shakes off its natural condition to appear as a product of culture and at the same time as a producer of culture. Marcel Mauss describes this process through the adoption of "body' techniques", i.e. knowledge that grows within the body itself. With this concept, Mauss wanted to highlight that behaviour like sexuality, eating, gestures long considered innate are not natural but acquired. Thomas Csordas developed Mauss' theory and those of Merleau-Ponty and Bourdieu, and invites scholars to flank anthropology of the body with an anthropology from the body: a way to rethink anthropology's main theoretical instrument, culture and, as a consequence, its methodologies and aims. For Csordas, culture is built inter-subjectively by individuals through inter-corporeal experience. See Marcel Mauss, *Esquisse d'une théorie générale de la magie* (Paris: Presses Universitaires de France, 1902); Thomas J. Csordas, "Incorporazione e fenomenologia culturale", *Annuario di Antropologia*, 3 (2003), 19–42; Ivo Quaranta, "Thomas Csordas: il paradigma dell'incorporazione", in *Discorsi sugli uomini. Prospettive antropologiche contemporanee*, ed. by Vincenzo Matera (Turin: Utet, 2008), pp. 49–71.

35   See Jack Goody, *Capitalism and Modernity: The Great Debate* (Cambridge: Polity Press, 2004); quote from the Italian edition, *Capitalismo e modernità. Il grande dibattito* (Milan: Cortina, 2005), p. 14.

These traces are not accepted by each individual in the same way, so there are continual variations on the theme. Therefore, it is typical of cultural dynamics to possess a certain level of diversity, because all cultural processes are potentially able to contain expressions of others.

If we look at different communities, social groups or some nation-states until just recently, it is possible to observe how people gathered around the expression of several cultural traits (religion, language, value systems). Whilst individuals share these traits, they are not totally identifiable with them: there is always a margin where members of a culture do not overlap because there are people of different ages, social classes and family backgrounds. Cultures do not have defined boundaries: they are the result of crossovers, diasporas and contacts, and did not develop within continuous historical processes.

It is sharing that allows the transmission of culture by ensuring that it "holds fast" as a system of signification, because in order to learn we must first be able to understand the teaching imparted. For this, a common code is necessary that makes the message we are sent semantically dense. This is not an arbitrary meaning but it is relevant to the context in which it is produced: it is a public, shared meaning.[36] I do not underestimate the theories of Hannerz when he questions the principle of sharing, that is to say, when he declares that in anthropology thinking of culture as something shared implies thinking of a homogeneous cultural distribution in society,[37] but I do think that sharing is attributable to an ample and sector-specific dimension, for example in certain classes or social groups.

On these grounds, I developed a methodological interpretation of the concept of culture that was useful during my fieldwork and data processing. A cultural sense is achieved when three closely interwoven dimensions come into play: language, memory and social relations. These dimensions were the *fil rouge* of my ethnographic method and link the testimonies I collected, bearing in mind that a possible over-schematisation of this viewpoint is only a hypothesis of illustration and as such certainly does not claim to be an exhaustive cultural analysis. These three aspects do not exist separately from one another: the development of linguistic codifications of collective memory and social

---

36   See Geertz (1973).
37   Hannerz (1996).

relations is a concurrent process, a form of training that occurs in unison. These three dimensions do, however, allow a basic level of analysis of human motivation, without prejudice to the investigation of other key social and cultural context variables like politics, economics and personal interests—although these variables are understood as events external to more profound dimensions, allowing us to understand the logic of their development.[38]

To document this interrelationship, it is necessary to record and transcribe the content of spoken testimonies (interviews, spontaneous tales, whispers and gestures) so as to enable their description and analysis. I attach great importance to language as a shared heritage in which, as Edward Sapir explains: "the mere fact of a common speech serves as a peculiarly potent symbol of the social solidarity of those who speak the language".[39] I also take into account the spatial and material context in which the *dabbawala* association works on a daily basis. The work of Mary Louise Pratt, which has been influential to my thinking, compares a language of community to a language of contact. She studies the functioning of language that crosses boundaries and pays attention to "zones of interaction" in which there are differences and inequalities.[40] The meaning given to language in my research, however, was broader than verbal communication, extending its semantic field to the body, facial expressions, proxemics and the ways of moving and dressing; all of which are, I believe, an integral part of the act of communication.

Formulating a description is not a neutral process. It is what Ward Goodenough defines as what we are able to build cognitively by observing phenomena.[41] This means that anthropologists describe on the basis of their own experience and, although they try to maintain an objective attitude to their research, they can only approximate the subject of their study. The description of cultural content can fall back on emic accounts to try to

---

38   Wagner (1981).

39   Edward Sapir, *Culture, Language and Personality: Selected Essays* (Berkeley: University of California Press, 1949).

40   Mary Louise Pratt, "Linguistic Utopias", in *The Linguistics of Writing: Arguments Between Language and Literature*, ed. by Nigel Fabb, Derek Attridge, Alan Durant and Colin MacCabe (Manchester: Manchester University Press, 1987), pp. 48–66. The bibliographical reference is in Ulf Hannerz, *Cultural Complexity: Studies in the Social Organization of Meaning* (New York: Columbia University Press, 1992).

41   Ward Goodenough, "Toward A Working Theory of Culture", in *Assessing Cultural Anthropology*, ed. by Robert Borofsky (New York: McGraw-Hill, 1994).

get closer to the way in which group members talk about their culture.[42] Goodenough's reflections are worth noting:

> An emic account lies more on the analysis of the patterns in people's opinions of people about what is or is not acceptable in specific situations and contexts than it does on their explanations or generalisations. An emic account, then, is a model of what one needs know to speak a language like a native speaker or to behave acceptably by the standards for socialised persons within a society. It is not merely an account of what they say about it.[43]

Thus the process is not a simple transcription and description of speech, but the reference to a diachronic dimension of the cultural experience that takes into account autobiographical and collective memory as a fundamental aspect of culture. Research into this aspect allows us to understand the social processes that permeate the cultural dimension, not in a linear fashion, but through overlaps, ruptures and reconstructions. It also allows us to work on the "threshold effect", mentioned by Norbert Elias, which can be profitably related to emotional, cognitive, behavioural and psychological processes.[44] It allows us to perceive in the subject's biography the passage through various experiences related to the "traces" mentioned by Goody, namely the transformations of the past recognisable in the present, and the innovations that each subject introduces, assimilates and transmits in the evolution of their generation. This is why individual and collective memories are co-present in the life of the subject who may not be fully aware of their predecessors' traces, but nonetheless experience and change them constantly. As Maurice Halbwachs puts it: "Our memories live in us as collective memories and we are reminded of them by others, even if we were the only ones involved in the events and only we saw the objects. The fact is that we are never really alone. There is no need for others to be

---

42   The emic-ethic conceptual pair was introduced by Kenneth L. Pike in the 1950s. Pike's terminology is based on a linguistic analogy and, in fact, "ethical" does not refer to ethics, the philosophy of judgment of human actions. In linguistics, language sounds can be described in two complementary perspectives: phonetic (hence "ethic"), which allows for an external description based on anatomical and physical parameters; the phonology perspective, whose basic unit is the phoneme (hence phonemic, hence "emic"), which is the minimum unit of a language sound distinguished by speakers of a particular language. This distinction made it possible to speak of significant and non-significant behaviour with regard to individuals who act. It is a distinction that has been greatly criticized, above all because it tends to identify a culture by its language and not the language as part of culture. See Alessandro Duranti, *Antropologia del linguaggio* (Rome: Meltemi, 2000).

43   Goodenough (1994), pp. 262–73.

44   Norbert Elias, *Über den Prozess der Zivilisation. I. Wandlungen des Verhaltens in den weltlichen Oberschichten des Abendlandes* (Basel: Verlag Haus zum Falken, 1939).

present, to be materially distinguishable from us because we all carry with us and within us a number of separate persons".[45]

The researcher may trace a multitude of people who have contributed to an interviewee's way of thinking, because collective memory exists in individual memory and collective representations are filtered through personal representation.[46]

The evolution of the concept leads us towards culture's relational dimension, confirming Geertz's analysis that culture arises and exists through human interaction. It is only through interaction that "people shape social structures and meanings in their contacts with one another [...]; and societies and cultures emerge and cohere as results of the accumulation and aggregation of these activities".[47] Goody goes on to say that culture is understood as the content of social relations.[48] Obvious as it may be, forms of cultural learning are developed from the small nucleus of the parenting circle and, in particular, the interaction with parents and siblings. Anthropologists have always studied family relations and their various forms worldwide. Beyond the differences, the relationship of deep affective involvement that exists between parents and children seems to be universal, even though management of emotions changes from context to context. This primary model of interaction will feature massively in the life of every generation, who will then perpetuate or change it, and in the latter case will perhaps succeed only in part.

---

45   Maurice Halbwachs, *Les cadres sociaux de la mémoire* (Paris: Presses Universitaires de France, 1952), it. trans., *La memoria collettiva* (Milan: Unicopli, 1987), p. 38.

46   George E. Marcus, "After the Critique of Ethnography: Faith, Hope and Charity, but the Greatest of these is Charity", in *Assessing Cultural Anthropology*, ed. by Robert Borofsky (New York: McGraw-Hill, 1994), pp. 40–53. I think it is worthwhile to note the inability to say everything in a testimony, that it can be comprehensive in giving a complete picture of the respondent's experience, and to take into account forms of oblivion, or what is forgotten, or what is unconsciously or consciously concealed by the subject in an interview. See Marc Augé, *Les Formes de l'oubli* (Paris: Payot et Rivages, 1998).

47   Ulf Hannerz, *Cultural Complexity. Studies in the Social Organization of Meaning* (New York: Columbia University Press, 1992) p. 15.

48   It is worth pointing out that I do not see the terms "social" and "cultural" in opposition to one another. As Jack Goody says, Geertz distinguishes between social structure and culture, meaning the first to be the "ongoing process of interactive behaviour" and the second "the complex of beliefs in terms of which individuals define their world, express their feelings and make their judgments." Geertz quoted by Jack Goody, "Culture and its Boundaries: A European View", in *Assessing Cultural Anthropology*, ed. by Robert Borofsky (New York: McGraw-Hill, 1994), pp. 250–61. I agree with Goody when he says that it is difficult to distinguish between the two levels since they interact incessantly and, in fact, social action would be meaningless without cultural action. There is no dichotomous relationship between content and interaction, because there would be a shift from the order of reality to that of the ideal, or from the material to the immaterial, in a continuous binary distinction.

It is within the family context, or within the institutions designed to stand in its place, that the fundamental forms of relational participation develop for each society. These forms are closely related to the expressive forms of verbal and non-verbal language, and to memory. Relational learning is gradual and the subject must be exposed repeatedly in order to incorporate knowledge models or patterns. These patterns are not rigid but provide flexible responses for the different situations in which the subjects find themselves. Each context reflects the hierarchical social order in which it exists—which is to say the production and reproduction of cultural forms related to power, class, social status, rank, gender and age inherent to historical ideologies.[49] In order to analyse the symbolic value of these relations, it is necessary to portray the full range of possible variations that these relations assume, and the power structures from which they spring. This methodological interpretation of culture makes the symbolic meanings of human action "visible".[50]

The ethnography of cultural diversities

In my effort to propose an ethnography of diversities I had to take into account the aforementioned dynamics which can be summarised as a few essential methodological points:[51]

- refocusing the discussion on concepts of culture and identity as unitary concepts underpinning a social group. We must define them in the plural and see cultures and identities as multiples.

---

49   See Erving Goffman, *Interaction Ritual: Essays on Face-to-Face Behavior* (New York: Anchor Books, 1967). See also Erving Goffman, *Frame Analysis: An Essay on the Organization of Experience* (Cambridge, MA: Harvard University Press, 1974).

50   Wagner (1981).

51   I have drawn extensively from Enzo Colombo, "Etnografia dei mondi contemporanei. Limiti e potenzialità del metodo etnografico nell'analisi della complessità", *Rassegna Italiana di sociologia*, 42:2 (2001), 205–30 (pp. 212–13). Pier Paolo Giglioli describes ethnography as "a style of quality research based on prolonged, direct observation whose aim is to describe and explain the significance of the practices of social players". See Pier Paolo Giglioli, "Una nuova rivista", *Etnografia e ricerca qualitativa*, 1 (2008), 3–8 (p. 4). Ethnography is fieldwork that allows an extended encounter between the researcher and the subjects, the spaces and the objects related to the research, with a common language whose scope is to enhance a plurality of points of view using a narrative style that counts among its main tools the unstructured interview, collection of biographies, participative observation and the use of videos as support documentation. See Ugo Fabietti and Vincenzo Matera, *Etnografia. Scritture e rappresentazioni dell'antropologia* (Rome: Carocci, 1997); A. Dal Lago and Rocco De Biasi (eds.), *Un certo sguardo. Introduzione all'etnografia sociale* (Rome: Laterza, 2002); and Antonio De Lauri and Luigi Achilli (eds.), *Pratiche e politiche dell'etnografia* (Rome: Meltemi, 2008).

Ethnography must necessarily include the strategies of construction of this multiplicity, the associated language and historical processes that are the basis of its construction;

- abandoning the idea that there are social groups and self-sufficient, culturally homogeneous places. The core of ethnographic work is increasingly realised in transit and frontier areas, in the study of cultural barrier mobility and of the cultures themselves;

- modifying the concept of observation: ethnographers are not strangers to the context studied and their action transforms their cultural experience and that of the interviewees, thus defining ethnography as a transforming practice and one that draws into the debate an ethnographer's totality of expression: senses, body, mind and soul;[52]

- modifying the concept of ethnographic description, since the ethnographer is unable to interpret and report completely the thoughts and actions of the interviewees. The ethnographer can only give one possible interpretation of the reality observed. This is why ethnography requires cumulative research and collaboration. Group work is much more in line with the complexity of cultural contexts and the ethnographic narrative is enriched by adverse set of viewpoints; and

- presuming the need for various professions in the research work and constant interaction with its participants, the illustrative models require two reflective dimensions: one for the players who are the subject of the research and one for the researchers to observe themselves. In particular, for the latter, it is crucial to think about the inevitable conditioning that the researcher brings with them, and the methods for accessing research data.

Alongside these provisions for methodology, the often dichotomous dynamics of expression inherent to globalisation processes must be taken into account, and may be summarised using Anthony McGrew's five binary oppositions:[53]

- *universalisation versus particularisation*: if, on one hand, globalisation universalises, in a manner of speaking, the central aspects and

---

52 On ethnography as a transformative practice, see Leonardo Piasere, *L'etnografo imperfetto. Esperienza e cognizione in antropologia* (Rome: Laterza, 2002).
53 The binary oppositions were suggested by Anthony McGrew, "A Global Society", in *Modernity and its Futures*, ed. by Stuart Hall, David Held and Anthony McGrew (Cambridge: Polity Press, 1992), pp. 62–113.

institutions of modern life, on the other it encourages ethnic and cultural particularisation through the exaltation of difference and local identities;

- *homogenisation versus differentiation*: extending the process of globalisation across the globe tends towards cultural homogenisation, but inevitably involves the assimilation of global into local parameters and thus the incessant production of "differences" or new localisms;

- *integration versus fragmentation*: globalisation creates new forms of organisation and transnational, regional or global communities on one hand, and on the other divides and fragments existing communities, both inside and outside nation-states;

- *centralisation versus decentralisation*: globalisation tends to concentrate power, knowledge, wealth, authority and institutions, while at the same time encouraging resistance movements and therefore the decentralisation of resources;

- *juxtaposition versus syncretisation*: by overlapping or bringing together different lifestyles, different cultures and social practices, globalisation may reinforce boundaries and cultural biases amongst groups, while also giving rise to hybrid, syncretic, or socially shared practices, ideas and values.

Practices: research considerations, themes and the transcription and editing of interviews

When I arrived in Mumbai I had to the face the daunting task of undertaking fieldwork in a large, unknown city. Fortunately, my first encounter with the president and the secretary of the *dabbawala* association, Raghunath Medge and Gangarama Talekar, had taken place earlier in Turin, at the 2006 Terra Madre [Slow Food] event organised every two years to promote an awareness of the world's various "food communities".[54] During that meeting I expressed my desire to carry out research into their association in Mumbai. They agreed, and so when I arrived in India, I already had my first interviews in place. I began to observe the daily work of meal

---

54   From the Terra Madre website: "Food communities are groups of persons who produce, process and distribute food of a good standard in a sustainable manner and are closely connected to the historical, social and cultural perspectives of a district. The communities share problems generated by intensive agriculture that damages natural resources and a mass food industry that aims to regiment tastes, placing at risk the very existence of small productions". See http://www.terramadre.info/pagine/rete [accessed 30 July 2012].

distribution and subsequently perform field investigation with in-depth interviews not only with Medge and Talekar but also with other *dabbawalas* at different levels of the organisation. The informed consent given by the association was not of secondary importance, because interviewees do not talk spontaneously. My search was immediately subject to supervision by the interviewees themselves: they watched my work and my behaviour. I never believed in pretending to bond with the subjects I studied. The *dabbawala* association members are almost all men and in a society where gender differences are marked, I tried to study their work in accordance with their customs.[55]

The course of the research was never as straightforward as it appears in written form. I tried to give a conversational, narrative slant to my exchanges with the *dabbawalas*, while keeping to the framework of issues I wanted to explore. I suspect I often made mistakes in how I posed questions to my interviewees—perhaps being indiscreet or too direct—but I think some of my own errors served as a stimulus, leading them to offer me a better understanding of the association's dynamics, their daily work, their faith and their relationship with Mumbai. Often I began with the biography of the interviewees, a simple way to open a conversation, setting both them and me at ease.[56]

In this first phase of work in Mumbai I had the help of the interpreter Francesca Caccamo, who undertook both translation and mediation. Together we discussed at length how, where and when to ask questions, but in practice we were far less rigid; so any place, situation and time was useful for observing and interacting with the *dabbawalas*. We worked in Mumbai's noisy streets; the association's offices; on trains when the *dabbas* were being delivered; and at road junctions. In ethnographic research there are no fixed protocols that can be used in all situations; rather, on each occasion I had to perceive the best survey strategy to use. Living with this uncertainty during field research is particularly difficult for any scholar, although I think this is precisely one of the great merits of ethnography: keeping our interpretation mechanisms alert and tuning into the view that our subject suggests to us, *together* we understand institutions, places, different customs and their meanings.

---

55 I must admit that situations arose which I found it particularly difficult to accept, for instance only eating after all the men had consumed their meal.

56 The bibliography of the collection of biographies and life stories is vast. See Rita Bichi, *L'intervista biografica. Una proposta metodologica* (Milan: Vita e Pensiero, 2002); and Luisa Passerini, *Storia e soggettività. Le fonti orali, la memoria* (Florence: La Nuova Italia, 1988).

During interviews and observation I first tried to develop an in-depth knowledge of the association's history, the founder's role, and the profiles of the customers in chronological order, in order to reconstruct the social, economic and gastronomic changes in Mumbai. From the outset the theme of the *varkari sampradaya*—the *dabbawala* faith—was clearly the ethical and symbolic backdrop to their work, and closely tied to their villages of origin. I began to understand the architecture of the association, its organisation chart and meal delivery system, and the roles and the logic of the underlying management. It became clear that food assumes a pivotal role in a diverse metropolis like Mumbai. The *dabbawala* delivery service has opened a tiny gap through which we achieve a better understanding of the many facets that food assumes in a context of major urban and cultural transformation.

Concomitantly with the field research, we undertook the long, patient work of transcription, translation and reworking of the interviews. Hindi interviews were transcribed by Usman Sheikh, a young actor who also helped in the translation of documents from Hindi. We tried to preserve the grammatical structures of Hindi so in the transcripts there will often be repetitions of phrases that may seem ungrammatical in another language. I performed the second stage of the work in Milan. I extrapolated the main themes that emerged from the interviews, eliminating the questions I asked the interviewees. From a dialogical, polyphonic approach we switched to individual narrative that took the form of short stories. In actual fact, the interviewer never disappears from the text because the triggers primed during the interview are the result of constant interaction and in this sense the questions merely start the reflection that serves to bring thoughts and memories to the surface.

In editing the texts I have tried to ensure that anecdotes, allegories and images remain true to the oral version. The same applies to autobiographical narrative because, if it is true, as Dennis Tedlock says, that telling a story is a matter of invention and imagination, it is equally true that it has to do with the effects of reality and "in seeking to create an appearance of reality a narrator has recourse to a number of devices [...]: gesture [...], quotation, onomatopoeia".[57] I also tried to achieve greater intelligibility for the reader, who may find it difficult to get their bearings in a transcript that sticks too closely to the spoken word, although I do warn—again quoting Tedlock— of the "opacity in the relationship between speech and writing".[58]

---

57   Dennis Tedlock, *The Spoken Word and the Work of Interpretation* (Philadelphia: University of Pennsylvania Press, 1983), p. 166

58   Tedlock (1983). These considerations come from research conducted in 2004 on the world of the Milan Stock Exchange. The research was coordinated by Roberta Garruccio

Alessandra Consolaro, a professor of Hindi language and literature at Turin University helped me to draw special attention to the "narrative character of cultural representations, to the stories built into the representational itself" while maintaining the freshness of the oral method and the rigour of grammatical understanding.[59] I tried not to fall into the trap of literal transcription and the myth of "structural nostalgia", which identifies a time before time, when narratives could not be corrupted by the manipulations of those who had participated in their actual collection.[60] This arrangement allows us to emphasise the creation and changing aspects of human activities, thus recognising that the understanding of cultural phenomena is necessarily an ongoing process.[61] The work of transcription and translation, and the use of ethnographic materials were intended to give an account of this constant transformation.

For the English version, the translator Angela Arnone has taken account of this methodological framework: she sought to reproduce a lively and spontaneous conversational style.[62] Translation of informal speech is as challenging as that of formal language and in this book there was a clear demarcation between the conveying of the research results and the transposition of the *dabbawala* interviews into credible English, via the Italian translation. The aim was to respect the informant, their culture, and their beliefs without seeming condescending or producing an English text that jarred on the reader. This was achieved by first translating all the background material—so the translator was informed of both the *dabbawalas'* history and their current situation—then the formulation of a draft translation of the interviews. These translated texts were then reviewed some days later, so they could be read with a fresh eye, and

---

and was included in the volume, Roberta Garruccio (ed.), *Le grida. Memoria, epica, narrazione del parterre di Milano* (Soveria Mannelli: Rubbettino, 2005). Also from the research conducted with Germano Maifreda on employees tasked with training in large companies and Italian trade unions, see Germano Maifreda and Sara Roncaglia (eds.), *Narrare la formazione. Grande impresa e sindacato* (Milan: Guerini, 2005).

59 James Clifford, "On Ethnographic Allegory", in *Writing Culture: The Poetics and Politics of Ethnography*, ed. by James Clifford and George E. Marcus (Berkeley: University of California Press, 1986), pp. 98–121 (p. 100).

60 In this way we can avoid museumising knowledge taken to be a static nature of experience. See Andreas Huyssen, "En busca del tiempo futuro", *Puentes*, 1:2 (2000), 12–29. Michael Herzfeld's *Cultural Intimacy: Social Poetics in the Nation-State* (New York: Routledge, 1997) uses the term "structural nostalgia" to analyse structural nostalgia as a collective representation of an uncorrupted heavenly order.

61 Tedlock (1983).

62 The translator's approach was based on many years of experience in rendering spoken language, including as an interpreter, and on the most underestimated yet most important of translation tools: intuition.

spoken aloud to a native English speaker to confirm that they sounded authentic.

Visual ethnography

During my fieldwork I used a small digital video camera to record the interviews, the meal delivery process, and the daily routine of the *dabbawalas*. At first I was concerned about the reaction of the interviewees: I was afraid I might disturb them with this invasive method of observation and interaction. This fear proved unfounded because of Mumbai's traditional relationship with visual culture. The city is the home of Bollywood and hosts the production of most Indian films, and its citizens have a relaxed attitude towards the camera. For example, I was once filming an interview in a lift with a female *dabbawala* who was taking a meal to a customer's office. I was doing my best not to film another person present in the lift without his consent, when a quite disappointed — and perhaps even offended — gentleman turned to me and asked: "Aren't you going to film me?"

In this way my research methodology was enhanced with visual ethnography as a tool for fieldwork collection and analysis.[63] Recent developments of this practice show two main technical and epistemological approaches. The first is the *positive* approach whereby the production of a film whose only aim is purely objective documentation, in its own perspective of so-called "salvage ethnography".[64] The shot material must be reprocessed and used as evidence in support of verbal reasoning.[65] The second is the *interpretative* approach, whereby the construction of the film

---

63   I use the term "visual ethnography" in relation to Francesco Faeta's definition when he writes "I believe that the study of audio-visual media and their use in anthropological research and recovery, or the study of optical recordings of certain cultural or social facts, should be defined properly as 'visual ethnography', while 'visual anthropology' should be considered the study of cultural productions that materialize images, viewing mechanisms and representation of a given human group, in relation to its broader social coordinates". See Francesco Faeta, *Strategie dell'occhio. Saggi di etnografia visiva* (Milan: Franco Angeli, 2003), pp. 10–11.

64   For "salvage ethnography" see Margaret Mead, "Visual Anthropology in a Discipline of Words", in *Principles of Visual Anthropology*, ed. by Paul Hockings (Berlin: Mouton de Gruyter, 1995), pp. 3–10. To investigate the history of this discipline, see Cecilia Pennacini, *Filmare le culture. Un'introduzione all'antropologia visiva* (Rome: Carocci, 2005); and Francesco Marano, *Camera etnografica. Storie e teorie di antropologia visuale* (Milan: Franco Angeli, 2007).

65   Paul Henley, "Film-making and Ethnographic Research", in *Image-based Research: A Sourcebook for Qualitative Researchers*, ed. by Jon Prosser (London: Routledge falmer, 2003), pp. 42–59.

is achieved by visual ethnographers following a storyline that emerges from the action. This perspective allows them to transmit the meaning of events being filmed from the perspective of the protagonists.[66] The two different approaches have taken the actual paths of anthropological epistemology divided between an aspiration towards the construction of a natural science that would turn their detached eye to the subject studied, and hermeneutic-interpretative approaches where there is the involvement of the scholar as well as an awareness of the cultural and social transformations triggered by the encounter between the researcher and the subject.

Over the last twenty-five years, the separation of those interested in the documentation and those interested in the documentary has not prevented ethnographic film from innovating both technically and stylistically.[67] These are innovations that I decided to bear in mind for my research, keeping shooting in a raw state and without stifling the technical devices that allowed me to produce the films. So the shots show all the persons who took part in the filming, the mediator, and my questions and considerations as I filmed. Nevertheless, my approach did not risk being ingenuous, thinking I would be able to use the images "simply" in a realistic key.[68]

In using video support I was able to obtain a dual benefit: on the one hand it provided me with a semantically richer, fuller documentation and storage of collected material, allowing me to be able to use the raw shot as the mirror for the written ethnographic record, in order to hear the complete oral account given by each interviewee and be able to compare it with the transformed interview texts included in the book. Moreover, the filmed ethnographic material is especially effective in representing the performative and symbolic aspects of a culture, showing contexts, gestures, and "know-how" for these members of society. Thus, I hope that the *dabbawala* delivery process, the signs on *dabbas*, the crowded trains, the noisy streets, the house-to-house pick up and affective, emotional and psychological aspects connected to their (and my) work are more understandable.

---

66 Ibid.
67 I recommend monitoring these aspects on http://www.visualanthropology.net.
68 On my part there was the awareness that each image was the result of a cultural construction that had to take into account, as Faeta writes: "links that connect visual national cultures, with their plots, [and] their social and political conditions" (2003), p. 8.

Glossary

Sanskrit terms in bold are given according to the IAST (International Alphabet of Sanskrit Transliteration) transcription standard. The main reference used for this glossary is the *Oxford Hindi-English Dictionary*, edited by R. S. McGregor (Oxford: Oxford University Press, 2007).

Agni	**Agni**: the two-headed god of fire and the recipient of daily sacrifice as messenger of the gods. One of his heads signifies immortality, while the other is considered a symbol of life renewal.
Ahimsa	**ahiṃsā**: "the avoidance of violence", a fundamental ethical virtue of Jainism, also respected in Buddhism and Hinduism.
Alandi	**Āḷandī**: a city and a municipal council in Pune district in the state of Maharashtra. Alandi is a place of pilgrimage and is venerated by many Hindus. A temple complex has been built near the spot of Sant Dnyaneshwar's *samadhi*. It is visited by thousands of pilgrims, in particular those of the Varkari sect.
Annadatta	**anna-dātā**: provider of food.
Annamaya Kosha	**anna-maya-kośa**: the "food-apparent-sheath", i.e. the physical body as a receptacle of nutrients in the Vedantic philosophy; the outer, less refined of the five illusory "sheaths" enveloping one's true self.
Annapurna Mahila Mandal	**Annapūrṇa Mahilā Maṇḍal**: an organisation working for the inn-runners since 1975 in Mumbai.

Artha	**artha**: purpose, motive, wealth, economy or gain. The term usually refers to the idea of material prosperity.
Atharva Veda	**Atharva-veda**: a sacred text of Hinduism and one of the four Vedas, often called the "fourth Veda".
Avatar	**avatār**: the incarnation of a Hindu deity in human or animal form.
Ayodhya	**Ayodhyā**: an ancient city of India adjacent to Faizabad city in Faizabad district of Uttar Pradesh. Ayodhya is a popular Hindu pilgrim centre, closely associated with Lord Ram, the seventh incarnation of Lord Vishnu. According to the *Ramayana*, the city was founded by Manu, the law-giver of the Hindus.
Babri	**Bābrī**: a mosque in Ayodhya on Ramkot Hill ("Rama's fort"). It was destroyed by Hindu pilgrims in 1992 in an outburst of anti-Islamic violence.
Balti	**bālṭī**: bucket.
Bhagavad Gita	**Bhagavad-gītā**: a 700-verse Dharmic scripture that is part of the ancient Sanskrit epic, one of the major works in the Indian literary tradition.
Bhagavān	**Bhagavān**: a term used to indicate the Supreme Being or Ultimate Truth in some traditions of Hinduism.
Bhai	**bhāī**: brother.
Bhaiya	**bhaiyā**: an endearing term for brother (i.e. "little brother"), also used for friends and acquaintances.
Bhakti	**Bhakti**: historical South Asian devotional movement, particularly active within Hinduism, that emphasises the love of a devotee for his or her personal god.
Bharatiya Janata Party (BNP)	**Bharatiyā Janatā Pārṭī**: a political party in India established in 1980.
Bhimashankar	**Bhīmaśaṅkar Temple**: located near Pune, at the source of the river Bhima.

Bhora	**Bohrā**: a modern Muslim Shiite sect of western India retaining some Hindu elements.
Bhumiputra	**bhūmi-putra**: "son of the soil", a term used by ethnic-Indians outside India, often with nativist and religious connotations.
Bombay	**Bambaī**: the capital city of the Indian state of Maharashtra, now known as Mumbai.
Brahma	**Brahmā**: in the Trimurti (the Hindu trinity), Brahma is the Immense Being, the orbiting force that creates space and time. Brahma is the point of equilibrium between Vishnu, his primarily creative, preserving force, and Shiva, the primarily destructive force. He represents the possibility of existence that arises when the opposing tendencies are coordinated.
Brahman	**brāhmaṇ**: a Hindu of the highest caste traditionally assigned to religious priesthood.
Chakra	**cakra**: "wheels", a term denoting any of several key points of physical or spiritual energy in the human body according to yoga philosophy.
Chawl	**cāḷ**: a large tenement house, found primarily in the factory cities of India.
Coolie	**kūlī**: an unskilled labourer or porter usually in or from the Far East hired for low or subsistence wages.
Dabba	**ḍabbā**: "box"; in Mumbai it is the word used for the multi-layered metal container used to transport prepared food by the *dabbawalas*.
Dabbawala	**ḍabbāvālā**: also spelled as *dabbawalla* or *dabbawallah*, the word literally means "box (-carrying) person", i.e. the "tiffin-box bearers" of Mumbai.
Dada	**dādā**: means elder brother in Bengali and means grandfather in Hindi.
Dalit	**dalit**: a group of people traditionally regarded as "untouchable". Dalits are a mixed population, consisting of numerous castes from all over South Asia.

Dhaba	**ḍhābā**: popular restaurants that generally serve local cuisine, and also act as truck stops.
Dharamshala	**dharmśālā**: an Indian religious guesthouse.
Dharma	**dharma**: "law", a concept of central importance in Indian philosophy and religion; the proper, pure way to be and act. It also indicates religious orthodoxy and orthopraxis in Hinduism.
Dholak	**ḍholak**: a South Asian two-headed hand-drum.
Diwali	**Dīvālī**: popularly known as the "festival of lights"; a five-day festival that celebrates the attainment of *nirvana* by the sage Mahavira in 527 BCE. In the Gregorian calendar, Diwali is celebrated between mid-October and mid-November.
Dosa	**dosā o ḍosā**: a fermented *crêpe* or pancake made from rice batter and black lentils.
Dvija	**dvija**: "twice-born", members of the first three higher *varnas*: Brahmins, Kshatriyas and Vaishyas.
Ekanath	**Ekanāth** (1533–1599): a prominent Marathi saint, scholar and religious poet.
Ghee	**ghī**: a kind of clarified semi-fluid butter, used especially in Indian cooking.
Gujarat	**Gujarāt**: a state in the Indian Union, located in north-western India.
Guna	**guṇa**: "string" or "a single thread or strand of a cord or twine"; by extension, it may mean "a subdivision, species, kind, quality", or an operational principle or tendency.
Guru	**guru**: a personal religious teacher and spiritual guide in Hinduism, but also a teacher and especially intellectual guide more generally.
Hindu	**hindū**: an adherent of Hinduism.

Hindutva	**hindūtva**: "Hinduness", a word coined by Vinayak Damodar Savarkar in his 1923 pamphlet *Hindutva: Who is a Hindu?*, used today to indicate the many facets of a political movement advocating Hindu nationalism and Hindu religious hegemony in India.
Holi:	**Holi**: a religious Spring festival celebrated in Hinduism as the "feast of colours".
Hyderabad	**Haidarābād**: the capital of the Indian state of Andhra Pradesh.
Izzat	**izzat**: the concept of honour prevalent in the culture of North India and Pakistan.
Jati	**jāti**: the caste system (literally "birth"). The term appears in almost all Indian languages and is related to the idea of lineage or kinship group.
Jejori	**Jejurī**: a city in Puṇe district.
Jnanadeva	**Jñānadeva** (1275–1296): a thirteenth-century Maharashtrian Hindu saint, poet and philosopher.
Jnaneshvari	**Jñāneśvarī**: a commentary on the *Bhagavad Gita* completed in 1290.
Jyoti Ling	**jyotirliṅg**: "Lingam (pillar) of light"; a sacred symbol that represents the permanent abode of Lord Shiva.
Kaccha	**kaccā**: uncooked.
Kaka	**kākā**: uncle.
Karma	**karma**: a concept which explains causality through a system of rebirth, where beneficial effects are derived from past beneficial actions and harmful effects from past harmful actions.
Khanawal	**khānāvālā**: small restaurants.

Khoja	**khojā**: a collective denomination for a group of diverse peoples — originally practitioners of Hinduism — originating from the Indian subcontinent. The word Khoja derives from Khwaja, a Persian/Turkic honorific title.
Kirtan	**kīrtan**: a kind of call-and-response chanting or "responsory" performed in India's devotional traditions.
Kolapur	**Kolhāpur**: a village in Radhanpur Taluk, Patan district, Gujarat.
Koli	**kolī**: historically, an Indo-Aryan ethnic group native to Rajasthan, Himachal Pradesh, Gujarat, Maharashtra, Uttar Pradesh and Haryana states.
Konkani	**koṇkaṇi**: an Indo-Aryan language in Western India.
Krishna	**Kṛṣṇa**: literally "black, dark blue", the name of a Hindu deity, an *avatar* of Vishnu. Krishna is often described and portrayed as an infant or young boy playing a flute, or as a youthful prince giving direction and guidance.
Kshatriya	**kṣatriya**: "warrior", one of the four traditional *varna*, or social orders in ancient India. Kshatriya constituted the military elite of the social system outlined by Hindu law.
Kunbi	**kumbī**: a prominent community of Karnataka. They can also be found in Tamil Nadu, Andhra Pradesh, Pondicherry, Karnataka, Kerala, Orissa and Maharashtra. Traditionally they belong to the fourth of the Hindu *varnas*, the so-called Sudra Kunbis.
Mahabharata	**Mahābhārata**: a major Sanskrit epic of ancient India, it contains an important conversation between the Pandava prince Arjuna and his guide Krishna on a variety of philosophical, spiritual and devotional subjects.
Maharashtra	**Mahārāṣṭra**: a state in Western India.

Mami	**māmī**: "mother's brother", maternal uncle.
Manas	**manas**: "mind", in the Hindu tradition it is the source of apparent reality (*maya*), the creator of everything we perceive as real.
Manomaya Kosha	**mano-maya-kośa**: "mind-stuff-apparent-sheath", one of the outer "sheaths" or illusory ("apparent") layers that enclose one's true self (or "soul"/*atman*) according to Vedantic philosophy.
Manu-Smrti	**Manu-smṛti**: "Laws of Manu"; the most authoritative of the books of the Hindu code (*Dharma-shastra*) in India. *Manu-smrti* is the popular name of the work, which is officially known as *Manava-dharma-shastra*. It is attributed to the legendary first man and lawgiver, Manu. In its present form, it dates from the first century BCE.
Maratha	**marāṭha**: an Indian warrior caste, found predominantly in the state of Maharashtra. The term Marāṭhā has two related usages: within the Marathi-speaking region it describes the dominant Maratha caste; historically, it describes the Maratha Empire founded by Shivaji in the seventeenth century.
Masala	**masālā**: a mixture of spices, popular in Indian cuisine.
Mleccha	**mlecch**: "foreigner", "non-Vedic barbarian"; a derogatory term used by native Indians in ancient India for any people of non-Indian extraction.
Moksha	**mokṣa**: "liberation", the final extrication of the soul from samsara and the bringing to an end of all the suffering involved in being subject to the cycle of repeated death and rebirth (reincarnation).
Mughal	**Mugal**: an Islamic empire set up in the Indian subcontinent by descendants of the Mongol conquerors of Asia from about 1526 to 1757.
Mukadam	**muqaddam**: "leader, chief, head", i.e. someone that is held in high regard and leads a group.

Mumbai	**Mumbaī**: capital of the State of Maharashtra; see *Bombay*.
Nadi	**nāḍī**: "channel", "stream", or "flow"; in yoga philosophy it refers to the channels of energy linking up the various chakra in the human body.
Namadeva	**Nāmadeva** (1270–1350): a poet and saint from the Varkari sect of Hinduism.
Narayan	**Nārāyaṇa**: another name for Vishnu; see *Satyanarayan*.
Pakka	**pakkā**: "cooked", also spelled *pukkah*, sometimes used to indicate genuineness or the "original, proper" way to do things.
Pandharpur	**Paṇḍharpur**: an important pilgrimage city on the banks of Bhimā river in Solāpur district, Maharashtra.
Parsee	**Pārsī**: a member of a group of followers in India of the Iranian prophet Zoroaster. The Parsees, also spelled Parsis, whose name means "Persians", are descended from Persian Zoroastrians who emigrated to India to avoid religious persecution by the Muslims.
Pav Bhaji	**pāv bhājī**: a fast food dish that originated in Marathi cuisine.
Prana	**prāṇa**: "exhalation of breath", i.e. life-force, vital energy.
Pranamaya kosha	**prāṇā-maya-kośa**: "air-apparent-sheath", one of the outer "sheaths" or illusory ("apparent") layers that enclose one's true self according to Vedantic philosophy.
Prasad	**prasād**: material substance that is first offered to a deity in Hinduism and then consumed.
Puja	**pūjā**: a religious ritual performed by Hindus as an offering to deities.
Pune	**Puṇe**: a city in Maharashtra state; Pune city is the capital of Pune district.
Purusha-Sukta	**Puruṣa-sūkta**: a hymn of the Rigveda, dedicated to the Purusha, the "Cosmic Being".

Rajasik	**rājas**: a class of foods that are bitter, sour, salty, pungent, hot, or dry, and are thought to promote sensuality, greed, jealousy, anger, delusion, and irreligious feelings in Ayurvedic philosophy.
Rajgurunagar	**Rājgurunagar**: a town in Pune district.
Rama	**Rāma**: Rama or Shri Ram (Lord Ram) is the seventh avatar of the god Vishnu.
Ramlila	**Rāmlīlā**: literally "Rama's play"; a dramatic folk re-enactment of the life of Lord Ram.
Rashtriya Swayamsevak Sangh (RSS)	**Rāṣṭrīya Svayaṁsevak Saṅgh**: the National Self-Service Organisation, a group founded in 1925 in opposition to Mohandas Gandhi and dedicated to the propagation of orthodox Hindu religious practices.
Rig-veda	**Ṛg-veda**: an ancient Indian sacred collection of Vedic Sanskrit hymns. It is counted among the four canonical Vedas.
Samsara	**saṃsāra**: the endless cycle of birth, suffering, death and rebirth.
Samskar	**saṃskāra**: a series of sacraments, sacrifices and rituals that serve as rites of passage and mark the various stages of human life.
Samyukta Maharashtra	**Saṃyukt Mahārāṣṭra**: "United Maharashtra Committee"; an organisation that spearheaded the demand in the 1950s for the creation of a separate Marathi-speaking state out of the (then-bilingual) state of Bombay in western India, with the city of Bombay as its capital.
Sattvici	**sāttvik**: a term denoting a class of foods that are fresh, juicy, light, nourishing, and tasty, and thus give necessary energy to the body and help achieve nutritional and energetic balance according to the Ayurvedic tradition.

Satyanarayan	**satya Nārāyaṇa**: a term referring to Shri Vishnu (Lord Vishnu) understood in his infinite and all-pervading form.
Shakti	**śakti**: the personification of divine feminine creative power and also a term for the manifestation of the creative principle (of a god or goddess). It is the primordial cosmic energy, the interplay of dynamic forces that are thought to move through the entire universe in Hinduism.
Shankara	**Ādi Śaṅkara** (circa 788–820 CE): philosopher and theologian, most renowned exponent of the Advaita Vedanta school of philosophy, from whose doctrines the main currents of modern Indian thought are derived. He wrote commentaries on the *Brahma-sutra*, the principal *Upanishads*, and the *Bhagavad Gita*, affirming his belief in one eternal unchanging reality (*brahman*) and the illusion of plurality and differentiation.
Shiv Sena	**Śiv-senā**: an Indian political organisation founded in 1966 by political cartoonist Bal Thackeray. The party originally emerged out of a movement in Mumbai demanding preferential treatment for Maharashtrians over migrants to the city.
Shiva	**Śiv/Rudra**: the Destroyer or the Transformer, a major and ancient Hindu deity. In the Trimurti, the Hindu trinity, he is darkness, the centrifugal force, dispersing and destroying all that exists. Shiva is the second of two interrelated and complementary tendencies. Each degree of manifestation of one is reversed with regard to the degree of manifestation of the other. If Vishnu is the centripetal force, Shiva represents the centrifugal: everything that has a beginning must end and Shiva presides over this passage through destruction and disintegration, the return to quiescence and sleep that occurs before any awakening or renewal is possible. As such, Shiva is regarded as a positive force, indispensable to sustain reality and its perpetual change.

Shudra	**śūdra**: the fourth and lowest of the traditional *varna*, or social classes, of India, traditionally artisans and labourers.
Solapur	**Solāpur**: a city in Maharashtra state on the Sina River.
Swastika	**svastik**: an equilateral cross with arms bent at right angles, all in the same rotary direction, usually clockwise. The swastika as a symbol of prosperity and good fortune is widely distributed throughout the ancient and modern world.
Tamasic	**tāmas**: a class of foods that are dry, old, foul, or unpalatable, and are thought to promote pessimism, ignorance, laziness, criminal tendencies, and doubt in the Ayurvedic tradition.
Tandoor	**Tandūr**: an Indian method of cooking over a charcoal fire in a *tandoor*, a cylindrical clay oven.
Tiffin	**ṭiphin**: Anglo-Indian term for "a light meal".
Topi	**ṭopī**: a white coloured sidecap, pointed in front and back and having a wide band, popular among the *dabbawalas*.
Trimurti	**Trimūrti**: the triad of gods consisting of Brahma the Creator, Vishnu the Preserver, and Shiva the Destroyer as the three highest manifestations of the one ultimate reality.
Tukaram	**Tukārām** (1608–1645): a prominent Varkari *sant* and spiritual poet during a Bhakti movement in India.
Udana	**uṛāna**: "upward moving air", i.e. the upward, transformative movement of the life-energy according to Vedantic philosophy. It governs growth of the body, the ability to stand, the powers of speech, the profusion of effort, enthusiasm and will.
Upajati	**upajāti**: "sub-class", or sub-caste.
Upanishad	**Upaniṣad**: a collection of philosophical texts which form the theoretical basis for the Hindu religion. They are also known as Vedanta.
Uttapam	**uttapam**: a thick pancake, made by cooking ingredients in a batter.

Vada pav	**vaṛā pāv**: a popular vegetarian fast food dish native to the Indian state of Maharashtra.
Vaishya	**vaiśya**: the third of four castes in Indian society, made up by merchants.
Varkari Sampradaya	**Vārkarī sampradāya**: a Vaishnava religious movement within the *bhakti* spiritual tradition of Hinduism, geographically associated with the Indian states of Maharashtra and northern Karnataka. Varkaris worship Vithoba (also known as Vitthal), the presiding deity of Pandharpur, regarded as a form of Krishna.
Varna	**varṇa**: "colour" (also: "shade, kind, quality"), used to define the four traditional castes of India in terms of social standing and economic function.
Veda	**Veda**: "knowledge", a corpus of religious and philosophic texts originating in ancient India. Composed in Vedic Sanskrit, the texts constitute the oldest layer of Sanskrit literature and the oldest scriptures of Hinduism.
Vijnanamaya Kosha	**vijñāna-maya-kośa**: "wisdom-apparent-sheath", one of the inner "sheaths" or illusory ("apparent") layers that enclose one's true self according to Vedantic philosophy.
Vishnu	**Viṣṇu**: Vishnu the Immanent is the centripetal force that creates light; Vishnu is the power of God through whom all things exist. Vishnu constantly re-invents the world and has become a symbol of continuity and eternal life.
Vishwa Hindu Parishad (VHP)	**Viśva Hindū Pariṣad**: a Hindu organisation founded in India in 1964 to protect, promote and propagate Hindu values of life.
Vithoba	**Viṭhobā**: a Hindu god, worshipped predominantly in the Indian states of Maharashtra, Karnataka, Goa, Andhra Pradesh and Tamil Nadu. He is a manifestation of the god Vishnu or his *avatar* Krishna.
Yoni	**yoni**: female reproductive organ.

Select Bibliography

Abbott, Dina, "Women's Home-Based Income-Generation as a Strategy towards Poverty Survival: Dynamics of the 'Khanawalli' (Mealmaking) Activity of Bombay" (unpublished PhD thesis, The Open University, 1994).

Achaya, K. T., "In India: civiltà pre-ariana e ariana" in *Storia e geografia dell'alimentazione*, ed. by Massimo Montanari and Françoise Sabban, 2 vols. (Turin: Utet, 2006), pp. 144-52.

—, *A Historical Dictionary of Indian Food* (New Delhi: Oxford University Press, 1998).

—, *Indian Food: A Historical Companion* (New Delhi: Oxford University Press, 1994).

Adiga, Aravind, *The White Tiger* (New York: Free Press, 2008).

Aime, Elena, *Breve storia del cinema indiano* (Turin: Lindau, 2005).

Aime, Marco, "Introduzione. Del dono e, in particolare, dell'obbligo di ricambiare i regali", in Marcel Mauss, *Saggio sul dono. Forma e motivo dello scambio nelle società arcaiche* (Turin: Einaudi, 2002), pp. i–xxviii.

—, *Eccessi di culture* (Turin: Einaudi, 2004).

—, *La casa di nessuno. I mercati in Africa occidentale* (Turin: Bollati Boringhieri, 2002).

Aklujkar, Vidyut, "Sharing the Divine Feast: Evolution of Food Metaphor in Marathi Sant Poetry", in *The Eternal Food: Gastronomic Ideas and Experiences of Hindus and Buddhists*, ed. by R. S. Khare (Albany: State University of New York Press, 1992), pp. 95–116.

Al-Biruni, *Alberuni's India: An Account of the Religion, Philosophy, Literature, Geography, Chronology, Astronomy, Customs, Laws, and Astrology of India about A.D. 1030*, trans. by E.C. Sachau (New Delhi: Munshiram Manoharlal, 1992).

Allovio, Stefano, "La 'vera carne' dei pigmei: parabole identitarie e strategie alimentari in Africa centrale", paper presented at the conference *Piatto pieno, piatto vuoto, prodotti locali e appetiti globali*, Università Statale, Milan, 2 April 2008.

Anderson, Benedict, *Imagined Communities: Reflections on the Origin and Spread of Nationalism*, rev. ed. (London: Verso, 1991).

Appadurai, Arjun, "Gastro-Politics in Hindu South Asia", *American Ethnologist*, 8 (1981), 494–511.

—, "How to Make a National Cuisine: Cookbooks in Contemporary India", *Comparative Studies in Society and History*, 30 (1988), 3–24.

—, "New Logics of Violence", *Seminar*, 503 (2001), available at http://www.india-seminar.com/2001/503/503%20arjun%20apadurai.htm [accessed 19 July 2012].

—, *Modernity at Large: Cultural Dimensions in Globalization* (Minneapolis: University of Minnesota Press, 1996).

Arens, William, *The Man-Eating Myth: Anthropology and Anthropophagy* (Oxford: Oxford University Press, 1980).

Armellini, Antonio, *L'elefante ha messo le ali. L'India del XXI secolo* (Milan: Egea, 2008).

Armstrong, John, "Archetypal Diasporas", in *Ethnicity*, ed. by John Hutchinson and Anthony D. Smith (Oxford: Oxford University Press, 1996), pp. 120–26.

Augé, Marc, *Les Formes de l'oubli* (Paris: Payot et Rivages, 1998).

Babb, Lawrence A., "The Food of the Gods in Chhattisgarh: Some Structural Features of Hindu Ritual", *Southwestern Journal of Anthropology*, 26 (1970), 287–304.

Bacchiega, Mario, *Il pasto sacro* (Padova: Cidema, 1971).

Balakrishnan, Natarajan and Chung-Piaw Teo, "Mumbai Tiffin (dabba) Express", University of Singapore, 2004, available at http://www.bschool.nus.edu.sg/staff/bizteocp/dabba.pdf [accessed 17 July 2012].

Barnard, Alan, *History and Theory in Anthropology* (Cambridge: Cambridge University Press, 2000).

Barnstone, Aliki (ed.), *The Shambhala Anthology of Women's Spiritual Poetry* (Boston: Shambhala, 2002).

Barth, Fredrik (ed.), *Ethnic Groups and Boundaries* (Oslo: Oslo University Press, 1969).

Bauman, Zygmunt, *Globalisation: The Human Consequences* (New York: Columbia University Press, 1998).

Baumann, Gerd, *The Multicultural Riddle: Rethinking National, Ethnic and Religious Identities* (London: Routledge, 1999).

Baviskar, Amita and Raka Ray (eds.), *Elite and Everyman: The Cultural Politics of the Indian Middle Classes* (New Delhi: Routledge, 2011).

Beck, Ulrich, *Was ist Globalisierung* (Berlin: Suhrkamp Verlag, 1997).

Behal, Rana Partap and Marcel Van Der Linden (eds.), *Coolies, Capital, and Colonialism: Studies in Indian Labour History* (Cambridge: Cambridge University Press, 2006).

Berger, John, *Ways of Seeing* (London: Penguin, 1972).

Béteille, André, *Society and Politics in India: Essays in a Comparative Perspective* (New Delhi: Oxford University Press, 1992).

Bhabha, Homi K., *The Location of Culture* (London: Routledge, 1994).

Bhonsle, R. N., *Clerks in the City of Bombay* (unpublished MA thesis, University of Bombay, 1938).

Bichi, Rita, *L'intervista biografica. Una proposta metodologica* (Milan: Vita e Pensiero, 2002).

Bonacich, Edna, "A Theory of Middlemen Minorities", *American Sociological Review*, 38 (1973), 583–94.

—, "Middleman Minorities and Advanced Capitalism", *Ethnic Groups*, 2 (1980), 311–20.

Bonte, Pierre and Michel Izard (eds.), *Dictionnaire d'ethnologie et anthropologie* (Paris: Presses Universitaires de France, 1991).

Borja, Jordi and Manuel Castells, *Local and Global: The Management of Cities in the Information Age* (London: Earthscan, 1997).

Borsa, Giorgio, *La nascita del mondo moderno in Asia orientale. La penetrazione europea e la crisi delle società tradizionali in India, Cina e Giappone* (Milan: Rizzoli, 1977).

Boudan, Christian, *Géopolitique du goût* (Paris: Presses Universitaires de France, 2004).

Bourdieu, Pierre, *La distinction. Critique sociale du jugement* (Paris: Les Éditions de Minuit, 1979).

Braudel, Fernand, *Civilisation matérielle, économie et capitalisme, XV-XVIII siècle. Les structures du quotidien* (Paris: Armand Colin, 1979).

Breman, Jan, *At Work in the Informal Economy of India. A Perspective from the Bottom Up* (New Delhi: Oxford University Press, 2013).

—, *Outcast Labour in Asia: Circulation and Informalisation of the Workforce at the Bottom of the Economy* (New Delhi: Oxford University Press, 2010).

—, *Labour Migration and Rural Transformation in Colonial Asia* (Amsterdam: Free University Press, 1990).

—, *Of Peasants, Migrants and Paupers: Rural Labour Circulation and Capitalist Production in West Bengal* (Oxford: Oxford University Press, 1985).

Brown, Patricia, *Anglo-Indian Food and Customs* (New Delhi: Penguin, 1998).

Burnett-Hurst, Alexander Robert, *Labour and Housing in Bombay: A study in the Economic Condition of the Wage-earning Classes of Bombay* (London: King & Son, 1925).

Caccia, P., "La cucina nei libri. Brevissima storia dei ricettari di cucina italiani dalle origini ai giorni nostri", *Eat:ing*, September 2008, available at http://www.eat-ing.net/attach/lacucinaneilibri.pdf [accessed 18 July 2012].

Caillé, Alain, *Anthropologie du don. Le tiers paradigme* (Paris: Desclée de Brouwer, 1994), it. trans., *Il terzo paradigma. Antropologia filosofica del dono* (Turin: Bollati Boringhieri, 1998).

Callari Galli, Matilde, Mauro Ceruti and Telmo Pievani, *Pensare la diversità. Per un'educazione alla complessità umana* (Rome, Meltemi, 1998).

Camporesi, Piero, *Il paese della fame* (Milan: Garzanti, 2000).

Cardona, Giorgio Raimondo, *La foresta di piume. Manuale di etnoscienza* (Rome: Laterza, 1985).

Carter, Marina and Khal Torabully, *Coolitude: An Anthology of the Indian Labour Diaspora* (London: Anthem Press, 2002).

Castells, Manuel, *The Information Age: Economy, Society and Culture. The Rise of the Network Society*, vol. 1 (Oxford: Blackwell, 1996).

Castiglioni, Ida, *Dal multiculturalismo al diversity management. Una ricerca empirica sulla definizione e sulla misura della competenza interculturale nei servizi sanitari e sociali di Milano e della sua provincia* (Milan: Provincia di Milano, 2008), available at http://www.provincia.milano.it/export/sites/default/affari_sociali/Allegati/multiculturalismo_diversity.pdf [accessed 30 July 2012].

Cella Al-Chamali, Gabriella, *Ayurveda e salute. Come curarsi con l'antica medicina indiana* (Milan: Sonzogno, 1994).

Chakravarty, S., "Fast Food", Forbes Global, 8 October 1998.

Chandavarkar, Rajnarayan, The Origins of Industrial Capitalism in India: Business Strategies and the Working Classes in Bombay, 1900–1940 (Cambridge: Cambridge University Press, 1994).

Chandrasekhar, Ramasastry, Dabbawallahs of Mumbai, Richard Ivey School of Business, University of Western Ontario, 2004, available at http://beedie.sfu. ca/files/PDF/mba-new-student-portal/2011/MBA/Dabbawallahs_of_Mumbai_ (A).pdf [accessed 28 October 2012].

Chaudhuri, Kirti Narayan, Asia before Europe: Economy and Civilisation of the Indian Ocean from the Rise of Islam to 1750 (Cambridge: Cambridge University Press, 1991).

Chiva, Matty, Le doux et l'amer (Paris: Presses Universitaires de France, 1985).

Clifford, James, "On Ethnographic Allegory", in Writing Culture: The Poetics and Politics of Ethnography, ed. by James Clifford and George E. Marcus (Berkeley: University of California Press, 1986), pp. 98–121.

Cohen, Abner, Custom and Politics in Urban Africa: A Study of Hausa Migrants in Yoruba Towns, rev. ed. (London: Routledge, 2004).

Cologna, Daniele, Cina a Milano (Milan: Abitare Segesta, 2000).

— (ed.), Asia a Milano (Milan: Abitare Segesta, 2003).

Colombo, Enzo, "Etnografia dei mondi contemporanei. Limiti e potenzialità del metodo etnografico nell'analisi della complessità", Rassegna Italiana di sociologia, 42:2 (2001), 205–30.

—, Le società multiculturali (Rome: Carocci, 2002).

—, Gianmarco Navarini and Giovanni Semi, "I contorni del cibo etnico", in Cibo, cultura, identità, ed. by Federico Neresini and Valentina Rettore (Rome: Carocci, 2008), pp. 78–96.

— and Giovanni Semi (eds.), Multiculturalismo quotidiano. La pratica della differenza (Milan: Franco Angeli, 2007).

Conlon, Frank F., "Dining Out in Bombay/Mumbai: An Exploration of an Indian City's Public Culture", in Urban Studies, ed. by Sujata Patel and Kushal Deb (New Delhi: Oxford University Press, 2006), pp. 390–413.

Connerton, Paul, How Societies Remember (Cambridge: Cambridge University Press, 1989).

Corbin, Alain, Le miasme et la jonquille. L'odorat et l'imaginaire social aux XVIIIè et XIXè siècles (Paris: Flammarion, 2005).

Coser, Rose Laub, In Defense of Modernity: Role Complexity and Individual Autonomy (Stanford, CA: Stanford University Press, 1991).

Crèvecoeur, J. Hector St John de, Letters From an American Farmer (New York: Fox, Duffield, 1904), available at http://xroads.virginia.edu/~HYPER/crev/home. html [accessed 30 July 2012].

Csordas, Thomas J., "Incorporazione e fenomenologia culturale", Annuario di Antropologia, 3 (2003), 19–42.

Cuche, Denys, La notion de culture dans les sciences sociales (Paris: Éditions La Découverte, 2001).

Dal Lago, Alessandro, *Non-Persone. L'esclusione dei migranti in una società globale* (Milan: Feltrinelli, 1999).

— and De Biasi Rocco (eds.), *Un certo sguardo. Introduzione all'etnografia sociale* (Rome: Laterza, 2002).

Dallapiccola, Anna L., *Dictionary of Hindu Lore and Legend* (London: Thames and Hudson, 2002).

Daniélou, Alain, *Mythes et Dieux de l'Inde. Le polythéisme hindou* (Paris: Editions du Rocher, 1992).

Das, N. K., "Cultural Diversity, Religious Syncretism and People of India: An Anthropological Interpretation", *Bangladesh e-Journal of Sociology*, 3:2 (2006), available at http://www.bangladeshsociology.org/BEJS%203.2%20Das.pdf [accessed 17 July 2012].

Davis, Kingsley, *The Population of India and Pakistan* (Princeton: Princeton University, 1951).

De Lauri, Antonio and Luigi Achilli (eds.), *Pratiche e politiche dell'etnografia* (Rome: Meltemi, 2008).

Della Casa, Carlo, *Il Gianismo* (Turin: Bollati Boringhieri, 1993).

Demel, Walter, *Wie die Chinesen gelb wurden. Ein Beitrag zur Frühgeschichte der Rassentheorien* (Bamberg: Förderverein Forschungsstiftung Überseegeschichte, 1993).

Desai, Urmila, *The Ayurvedic Cookbook* (New Delhi: Motilal Banarsidass, 1994).

Doniger, Wendy (ed.), *The Laws of Manu* (London: Penguin, 1991).

Donner, Henrike, *Domestic Goddesses: Maternity, Globalization and Middle-class Identity in Contemporary India* (London: Ashgate, 2008).

Dossal, Mariam, *Theatre of Conflict, City of Hope. Bombay/Mumbai, 1660 to Present Times* (New Delhi: Oxford University Press, 2010).

Douglas, Mary, *Implicit Meanings* (London: Routledge, 1975).

—, *Purity and Danger: An Analysis of Concepts of Pollution and Taboo* (London: Routledge, 1970).

Drèze, Jean and Amartya Sen, *Hunger and Public Action* (New Delhi: Oxford University Press, 1989).

Dumont, Louis, *Homo hierarchicus. Le système des castes et ses implications* (Paris: Gallimard, 1966).

Dupront, Alphonse, *L'Acculturazione. Per un nuovo rapporto tra ricerca storica e scienze umane* (Turin: Einaudi, 1966).

Duranti, Alessandro, *Antropologia del linguaggio* (Rome: Meltemi, 2000).

Dwyer, Rachel and Christopher Pinney, *Pleasure and the Nation: The History, Politics and Consumption of Public Culture in India* (New Delhi: Oxford University Press, 2001).

Ehrenreich, Barbara and Arlie Russell Hochschild, *Global Woman: Nannies, Maids, and Sex Workers in the New Economy* (London: Granta, 2003).

Eliade, Mircea, *Techniques du Yoga* (Paris: Gallimard, 1948).

—, *Yoga, Immortality and Freedom* (London: Routledge, 1958).

Elias, Norbert, *Über den Prozess der Zivilisation. I. Wandlungen des Verhaltens in den weltlichen Oberschichten des Abendlandes* (Basel: Verlag Haus zum Falken, 1939).

Enthoven, Reginald E., "Kojah", in *The Tribes and Castes of Bombay*, ed. by Reginal E. Enthoven, 3 vols. (Bombay: Government Central Press, 1921), vol. 2, pp. 218-30.

Epstein, A. L., *Ethos and Identity: Three Studies in Ethnicity* (London: Tavistock, 1981).

Fabietti, Ugo, "Diversità delle culture e disagio della contemporaneità", *I Quaderni del CREAM*, 9 (2009), available at http://www.unimib.it/upload/aa_fabietti.pdf [accessed 30 July 2012].

—, *L'identità etnica. Storia e critica di un concetto equivoco* (Rome: NIS, 1995).

—, *Storia dell'antropologia* (Bologna: Zanichelli, 1991).

— and Francesco Remotti (eds.), *Dizionario di antropologia* (Bologna: Zanichelli, 1997).

— and Vincenzo Matera, *Etnografia. Scritture e rappresentazioni dell'antropologia* (Rome: Carocci, 1997).

Faeta, Francesco, *Strategie dell'occhio. Saggi di etnografiavisiva* (Milan: Franco Angeli, 2003).

Falzon, Mark-Anthony, *Cosmopolitan Connections: The Sindhi Diaspora, 1860–2000* (New Delhi: Oxford University Press, 2005).

Fay, Peter Ward, *The Opium War 1840–1842* (Chapel Hill: University of North Carolina Press, 1975).

Featherstone, Mike (ed.), *Global Culture* (London: Sage, 1990).

Feldhaus, Anne, *Connected Places: Region, Pilgrimage, and Geographical Imagination in India* (New York: Palgrave Macmillan, 2003).

Ferrozzi, Claudio and Roy Shapiro, *Dalla logistica al Supply Chain Management. Teorie ed esperienze* (Turin: Isedi, 2000).

Filoramo, Giovanni (ed.), *Storia delle religioni IV: Religioni dell'India e dell'Estremo Oriente* (Rome: Laterza, 1996).

Fiorani, Eleonora, *Selvaggio e domestico. Tra antropologia, ecologia ed estetica* (Padua: Muzzio, 1993).

Fischer, Michael M. J., "Ethnicity and the Post-Modern Arts of Memory", in *Writing Culture: The Poetics and Politics of Ethnography*, ed. by James Clifford and George E. Marcus (Berkeley: University of California Press, 1984), pp. 194–233.

Fischler, Claude, *L'Homnivore* (Paris: Odile Jacob, 1990).

Flood, Gavin, *An Introduction to Hinduism* (Cambridge: Cambridge University Press, 1996).

Foglio, Antonio, *Il marketing agroalimentare. Mercato e strategie di commercializzazione* (Milan: Franco Angeli, 2007).

Ford-Grabowsky, Mary, *Sacred Voices: Essential Women's Wisdom through the Ages* (New York: HarperCollins, 2002).

Foucault, Michel, *Naissance de la biopolitique. Cours au collège de France, 1978–1979* (Paris: Gallimard-Seuil, 2004), it. trans., *Nascita della biopolitica. Corso al Collége de France, 1978–1979* (Milan: Feltrinelli, 2005).

Frawley, David, *Yoga and Ayurveda: Self-Healing and Self-Realisation* (Delhi: Motilal Banarsidass, 2000).

Frazer, James George, *The Golden Bough: A Study in Magic and Religion* (New York: Macmillan, 1922).

Freed, Stanley A., "Caste Ranking and the Exchange of Food and Water in a North Indian Village", *Anthropological Quarterly*, 43 (1970), 1–13.

Fyzee, Asaf A. A., "Bohoras", in *Encyclopaedia of Islam*, 12 vols. (Leiden: E. J. Brill, 1960–2005), vol. 1, pp. 1254-55.

Gallissot, René, Mondher Kilani and Annamaria Rivera (eds.), *L'imbroglio etnico in quattordici parole-chiave* (Bari: Dedalo, 2001).

Ganguly-Scrase, Ruchira and Timothy J. Scrase, *Globalisation and the Middle Classes in India: The Social and Cultural Impact of Neoliberal Reforms* (London: Routledge, 2008).

Garruccio, Roberta (ed.), *Le grida. Memoria, epica, narrazione del parterre di Milano* (Soveria Mannelli: Rubbettino, 2005).

—, *Minoranze in affari. La formazione di un banchiere: Otto Joel* (Soveria Mannelli: Rubbettino, 2002).

Geertz, Clifford, "The Uses of Diversity", in *Tanner Lectures on Human Values*, ed. by Sterling M. McMurrin, vol. 7 (Salt Lake City, UT: University of Utah Press, 1986), pp. 253-75.

—, *Interpretation of Cultures* (New York: Basic Books, 1973).

Giddens, Anthony, *The Consequences of Modernity* (Stanford, CA: Stanford University Press, 1990).

Giglioli, Pier Paolo, "Una nuova rivista", *Etnografia e ricerca qualitativa*, 1 (2008), 3–8.

Glazer, Nathan, *We Are All Multiculturalists Now* (Cambridge, MA: Harvard University Press, 1998).

— and Daniel P. Moynihan, "Why Ethnicity?", *Commentary*, 58 (1974), 33–39.

Godbout, Jacques T., *L'esprit du don* (Paris: Editions la découverte, 1992), it. trans. *Lo spirito del dono* (Turin: Bollati Boringhieri, 2002).

Godelier, Maurice, *L'énigme du don* (Paris: Fayard, 1996).

Goffman, Erving, *Frame Analysis: An Essay on the Organization of Experience* (Cambridge, MA: Harvard University Press, 1974).

—, *Interaction Ritual: Essays on Face-to-Face Behavior* (New York: Anchor Books, 1967).

Gombrich, Richard, *Theravada Buddhism: A Social History from Ancient Benares to Modern Colombo* (New York: Routledge, 1988).

Goodenough, Ward, "Toward A Working Theory of Culture", in *Assessing Cultural Anthropology*, ed. by Robert Borofsky (New York: McGraw-Hill, 1994), pp. 262–73.

Goody, Jack, "Culture and its Boundaries: A European View", in *Assessing Cultural Anthropology*, ed. by Robert Borofsky (New York: McGraw-Hill, 1994), pp. 250–61.

—, *Capitalism and Modernity: The Great Debate* (Cambridge: Polity Press, 2004), it. trans. *Capitalismo e modernità. Il grande dibattito* (Milan: Cortina, 2005).

Goswami, Manu, *Producing India: From Colonial Economy to National Space* (Chicago: University of Chicago Press, 2004).

Grendi, Edoardo (ed.), *L'antropologia economica* (Turin: Einaudi, 1972).

Grignaffini, Giorgio, "Estesia e discorsi sociali: per una sociosemiotica della degustazione del vino", in *Gusti e disgusti. Sociosemiotica del quotidiano*, ed. by Eric Landowski and José L. Fiorin (Turin: Testo & Immagine, 2000), pp. 214–32.

Guha, Ramachandra, *India After Gandhi: The History of the World's Largest Democracy* (London: Macmillan, 2007).

Guigoni, Alessandra (ed.), *Foodscapes. Stili, mode e culture del cibo oggi* (Monza: Polimetrica, 2004).

—, "L'alimentazione mediterranea tra locale e globale, tra passato e presente", in *Saperi e sapori del Mediterraneo*, ed. by Radhouan Ben Amara and Alessandra Guigoni (Cagliari: AM&D, 2006), pp. 81–92.

Gupta, Akhil and James Ferguson "Beyond 'Culture': Space, Identity, and the Politics of Difference", *Cultural Anthropology*, 7:1 (1992), 6–32.

Gupta, Dipankar (ed.), *Social Stratification* (New Delhi: Oxford University Press, 1991).

Gusman, Alessandro, *Antropologia dell'olfatto* (Rome: Laterza, 2004).

Gutmann, Amy (ed.), *Multiculturalism: Examining the Politics of Recognition* (Princeton, NJ: Princeton University Press, 1994).

Halbwachs, Maurice, *Les cadres sociaux de la mémoire* (Paris: Presses Universitaires de France, 1952), it. trans., *La memoria collettiva* (Milan: Unicopli, 1987).

Hannerz, Ulf, *Cultural Complexity: Studies in the Social Organization of Meaning* (New York: Columbia University Press, 1992).

—, *Transnational Connections: Culture, People, Places* (London: Routledge, 1996).

Hansen, Thomas Blom, *Wages of Violence: Naming and Identity in Postcolonial Bombay* (Princeton: Princeton University Press, 2001).

Harris, Marvin, *Good to Eat: Riddles of Food and Culture* (Long Grove, IL: Waveland, 1985).

Henley, Paul, "Film-making and Ethnographic Research", in *Image-based Research: A Sourcebook for Qualitative Researchers*, ed. by Jon Prosser (London: Routledge Falmer, 2003), pp. 42–59.

Héritier, Françoise, *Masculin-Féminin. La pensée de la différence* (Paris: Éditions Odile Jacob, 1996).

Herzfeld, Michael, *Cultural Intimacy: Social Poetics in the Nation-State* (New York: Routledge, 1997).

Hobsbawm, Eric and Terence Ranger (eds.), *The Invention of Tradition* (Cambridge: Cambridge University Press, 1983).

Hofstede, Geert and Gert J. Hofstede, *Cultures and Organizations: Software of the Mind*, 2nd ed. (New York: McGraw-Hill, 2005).

Huyssen, Andreas, "En busca del tiempo futuro", *Puentes*, 1:2 (2000), 12–29.

Inden, Ronald, *Imagining India* (Oxford: Blackwell, 1992).

International Commission for the Future of Food and Agriculture, *Manifesto on the Future of Food* (Florence: Arsia, 2006), available at http://commissionecibo.arsia. toscana.it/UserFiles/File/Commiss%20Intern%20Futuro%20Cibo/cibo_ing.pdf [accessed 19 July 2012].

Irudaya Rajan, Sebastian, *Catholics in Bombay: A Historical-Demographic Study of the Roman Catholic Population in the Archdiocese of Bombay* (Shillong: Vendrame Institute, 1993).

Iversen, Vegard and P. S. Raghavendra, "What the Signboard Hides: Food, Caste and Employability in Small South Indian Eating Places", *Contributions to Indian Sociology*, 40 (2006), 311–41. DOI: 10.1177/006996670604000302

Jacobs, Jane, *The Death and Life of Great American Cities* (New York: Random House, 1961).

Jaffrelot, Christoph and Peter van der Veer, *Patterns of Middle Class Consumption in India and China* (New Delhi: Sage, 2008).

Kakar, Sudhir and Katharina Kakar, *The Indians: Portrait of a People* (New Delhi: Penguin, 2007).

Kamat, Manjiri N., *Mumbai-Past and Present: Historical Perspectives and Contemporary Challenges* (Mumbai: Nehru Centre and Indus Source Books, 2013).

Karaka, Dosabhai Framji, *History of Parsees; Including their Manners, Customs, Religion and Present Position* (London: Macmillan, 1884).

Karve, Irawati, "On the Road: A Maharashtrian Pilgrimage", in *The Experience of Hinduism: Essays on Religion in Maharashtra*, ed. by Eleanor Zelliot and Maxine Berntsen (New York: State University of New York Press, 1988), pp. 142–73.

Katz, Nathan, *Who Are the Jews of India?* (Berkeley: University of California Press, 2000).

Khare, Ravindra S. (ed.), *Caste, Hierarchy and Individualism: Indian Critiques of Luis Dumont's Contributions* (New Delhi: Oxford University Press, 2006).

—, *The Eternal Food: Gastronomic Ideas and Experiences of Hindus and Buddhists*, ed. by R. S. Khare (Albany: State University of New York Press, 1992).

Kilani, Mondher, "L'ideologia dell'esclusione. Note su alcuni concetti-chiave", in *L'imbroglio etnico in quattordici parole-chiave*, ed. by René Gallissot, Mondher Kilani and Annamaria Rivera (Bari: Dedalo, 2001), pp. 9–36.

—, *L'Invention de l'autre. Essais sur le discours anthropologique* (Paris: Payot et Rivages, 1994).

King, Niloufer Ichaporia, *My Bombay Kitchen: Traditional and Modern Parsi Home Cooking* (Berkeley: University of California Press, 2007).

Kulke, Eckehard, *The Parsees in India: a Minority as an Agent of Social Change* (New Delhi: Vikas, 1978).

Kydd, J. C., "The First Indian Factories Act (Act XV of 1881)", *The Calcutta Review*, 293 (1918), 279–92.

Le Breton, David, *La Saveur du Monde: une anthropologie des sens* (Paris: Editions Métailié, 2006).

Le Guérer, Annick, *Les pouvoirs de l'odeur* (Paris: Odile Jacob, 1998).

Lévi-Strauss, Claude, "The Culinary Triangle", in *Food and Culture: A Reader*, ed. by Carole Counihan and Penny Van Esterik (New York: Routledge, 1997), pp. 36–43.

—, *L'Origine des manières de table* (Paris: Plon, 1968).

Lévi, Sylvain, *La Doctrine du sacrifice dans les Brahmanas* (Paris: Ernest Leroux, 1898).

Light, Ivan and Edna Bonacich, *Immigrant Entrepreneurs: Koreans in Los Angeles, 1965–1982* (Berkeley: University of California Press, 1988).

Livi Bacci, Massimo, *Popolazione e alimentazione. Saggio sulla storia demografica europea* (Bologna: Il Mulino, 1993).

Lutz, Catherine A. and Lila Abu-Lughod (eds.), *Language and the Politics of Emotion* (Cambridge: Cambridge University Press, 1990).

Machado, Felix, *Jnaneshvari: Path to Liberation* (Mumbai: Somaiya, 1998).

Mahajan, Gurpreet, "Indian Exceptionalism or Indian Model: Negotiating Cultural Diversity and Minority Rights in a Democratic Nation-State", in *Multiculturalism in Asia*, ed. by Will Kymlicka and Baogang He (Oxford: Oxford University Press, 2005), pp. 288–313.

Maifreda, Germano and Sara Roncaglia (eds.), *Narrare la formazione. Grande impresa e sindacato* (Milan: Guerini, 2005).

Malamoud, Charles, *Cooking the World: Ritual and Thought in Ancient India* (Delhi: Oxford University Press, 1996).

Mallik, U. K. and D. Mukherjee, "Sigma 6 Dabbawalas of Mumbai and their Operations Management: An Analysis", *The Management Accountant*, 42 (2007), 386–88.

Manekshaw, Bhicoo J., *Parsi Food and Customs* (New Delhi: Penguin, 1996).

Marano, Francesco, *Camera etnografica. Storie e teorie di antropologia visuale* (Milan: Franco Angeli, 2007).

Marcus, George E., "After the Critique of Ethnography: Faith, Hope and Charity, but the Greatest of these is Charity", in *Assessing Cultural Anthropology*, ed. by Robert Borofsky (New York: McGraw-Hill, 1994), p. 40–53.

—, and Michael M. J. Fischer, *Anthropology as Cultural Critique: An Experimental Moment in the Human Sciences* (Chicago: University of Chicago Press, 1986).

Marriot, McKim (ed.), *India through Hindu Categories* (New Delhi: Thousand Oaks; London: Sage, 1992).

—, "Caste Ranking and Food Transactions: A Matrix Analysis", in Milton Singer and Bernard S. Cohn, *Structure and Change in Indian Society* (Chicago: Aldine, 1968), pp. 133–72.

de Martino, Gianni, *Odori. Entrate in contatto con il quinto senso* (Melzo: Apogeo, 1997).

Masselos, Jim, "Defining Moments/Defining Events: Commonalities of Urban Life", in *Bombay and Mumbai: The City in Transition*, ed. by Sujata Patel and Jim Masselos (New Delhi: Oxford University Press, 2003), pp. 31–52.

—, *The City in Action: Bombay Struggles for Power* (New Delhi: Oxford University Press, 2007).

Massey, Douglas S., Joaquin Arango, Graeme Hugo, Ali Kouaouci, Adela Pellegrino and J. Edward Taylor, *Worlds in Motion: Understanding International Migration at the End of the Millennium* (Oxford: Oxford University Press, 1998).

Mauss, Marcel, *Esquisse d'une théorie générale de la magie* (Paris: Presses Universitaires de France, 1902).

—, *The Gift: The Form and the Reason for Exchange in Archaic Societies* (London: Routledge, 1990), it. trans., *Saggio sul dono. Forma e motivo dello scambio nelle società arcaiche* (Turin: Einaudi, 2002).

Mayer, Adrian C., *Caste and Kinship in Central India* (Berkeley: University of California Press, 1960).

Mayer, Hans, *Außenseiter* (Frankfurt am Main: Suhrkamp, 1975).

McGrew, Anthony, "A Global Society", in *Modernity and its Futures*, ed. by Stewart Hall, David Held and Anthony McGrew (Cambridge: Polity Press, 1992), pp. 62–113.

Mead, Margaret, "Visual Anthropology in a Discipline of Words", in *Principles of Visual Anthropology*, ed. by Paul Hockings (Berlin: Mouton de Gruyter, 1995), pp. 3–10.

Mehta, Suketu, *Maximum City: Bombay Lost and Found* (London: Headline Review, 2005).

Memmi, Albert, *Il razzismo* (Genoa: Costa e Nolan, 1989).

Menon, Meena and Neera Adarkar, *One Hundred Years One Hundred Voices: The Millworkers of Girangaon: An Oral History* (New Delhi: Seagull Books, 2004).

Miavaldi, Isabella, *La cucina ayurvedica* (Milan: Xenia, 1999).

Michaels, Walter Benn, *The Trouble with Diversity: How We Learned to Love Identity and Ignore Inequality* (New York: Metropolitan, 2006).

Minkowski, Eugène, *Vers une cosmologie. Fragments philosophiques* (Paris: Aubier-Montaigne, 1936), it. trans., *Verso una cosmologia* (Turin: Einaudi, 2005).

Mintz, Sidney W., *Sweetness and Power: The Place of Sugar in Modern History* (New York: Viking, 1985).

Montanari, Massimo and Françoise Sabban, *Storia e geografia dell'alimentazione*, 2 vols. (Turin: Utet, 2004).

Morin, Edgar, *Introduction à la pensée complexe* (Paris: ESF Éditeur, 1990), it. trans., *Introduzione al pensiero complesso* (Milan: Sperling and Kupfer, 1993).

Morris, Morris D., "The Emergence of an Industrial Labour Force in India", in *Social Stratification*, ed. by Dipankar Gupta (New Delhi: Oxford University Press, 1991), pp. 231–47.

Moulier Boutang, Yann, *De l'Esclavage au salariat: Economie historique du salariat bridé* (Paris: Presses Universitaires de France, 1998).

Muzzarelli, Maria Guiseppina and Fiorenza Tarozzi, *Donne e cibo. Una relazione nella storia* (Milan: Bruno Mondadori, 2003).

Nandy, Ashis, "The Twilight of Certitudes: Secularism, Hindu Nationalism and Other Masks of Deculturation", *Postcolonial Studies*, 1 (1998), 283–98.

Nestle, Marion, *Food Politics: How The Food Industry Influences Nutrition and Health* (Berkeley: University of California Press, 2002).

Nussbaum, Martha C., *The Clash Within: Democracy, Religious Violence, and India's Future* (Cambridge, MA: Harvard University Press, 2007).

Olivelle, Patrick, "Food in India", *Journal of Indian Philosophy*, 23 (1995), 367–80.

Pandit, Shrinivas, *Dabawalas* (New Delhi: Tata McGraw-Hill, 2007).

Panikkar, Raimon, *The Vedic Experience: Mantramanjari* (New Delhi: Sanctum, 1977).

Paolini, David, Tullio Seppilli and Alberto Sorbini, *Migrazioni e culture alimentari* (Foligno: Editoriale Umbra, 2002).

Parekh, C. S., *The Dabbawallas of Mumbai* (unpublished PhD thesis, Narsee Monjee College of Commerce and Economics, Mumbai, 2005).

Passerini, Luisa, *Storia e soggettività. Le fontiorali, la memoria* (Florence: La Nuova Italia, 1988).

Patel, Raj, *Stuffed and Starved: The Hidden Battle for The World Food System* (New York: Melville House, 2008).

Patel, Sujata and Alice Thoner (eds.), *Bombay: Mosaic of Modern Culture* (New Delhi: Oxford University Press, 1995).

— and Jim Masselos (eds.), *Bombay and Mumbai: The City in Transition* (New Delhi: Oxford University Press, 2003).

Pavanello, Deborah, *Cibo per l'anima. Il significato delle prescrizioni alimentari nella grandi religioni* (Rome: Mediterranee, 2006).

Pennacini, Cecilia, *Filmare le culture. Un'introduzione all'antropologia visiva* (Rome: Carocci, 2005).

Percot, Marie, "Dabbawalas, Tiffin Carriers of Mumbai: Answering a Need for Specific Catering", *HAL: Sciences de l'Hommeet de la Société*, 2005, available at http://halshs.archives-ouvertes.fr/docs/00/03/54/97/PDF/DABBA.pdf [accessed 10 August 2012].

Petrini, Carlo, *Buono, pulito e giusto. Principi di una nuova gastronomia* (Turin: Einaudi, 2005).

Piano, Stefano, "Lo hinduismo II. La prassi religiosa", in *Hinduismo*, ed. by Giovanni Filoramo (Rome: Laterza, 2002), pp. 171-246.

Piantelli, Mario, "Lo hinduismo. I. Testi e dottrine", in *Hinduismo*, ed. by Giovanni Filoramo (Rome: Laterza, 2002), pp. 61-170.

Piasere, Leonardo, *L'etnografo imperfetto. Esperienza e cognizione in antropologia* (Rome: Laterza, 2002).

Pirenne, Henri, *Les villes du moyen-âge. Essai d'histoire économique et sociale* (Brussels: Lamertin, 1927).

Plattner, Felix Alfred, *Christian India* (London: Thames and Hudson, 1957).

Portes, Alejandro (ed.), *The Economic Sociology of Immigration: Essays on Networks, Ethnicity, and Entrepreneurs* (New York: Russell Sage Foundation, 1995).

Poulain, Jean-Pierre, *Sociologies de l'alimentation, les mangeurs et l'espace social alimentaire* (Paris: Presses Universitaires de France, 2002).

Pratt, Mary Louise, "Linguistic Utopias", in *The Linguistics of Writing: Arguments Between Language and Literature*, ed. by Nigel Fabb, Derek Attridge, Alan Durant and Colin MacCabe (Manchester: Manchester University Press, 1987), pp. 48–66.

Punekar, Vinaja B., *The Son Kolis of Bombay* (Bombay: Popular Book Depot, 1959).

Quaranta, Ivo, "Thomas Csordas: il paradigma dell'incorporazione", in *Discorsi sugli uomini. Prospettive antropologiche contemporanee*, ed. by Vincenzo Matera (Turin: Utet, 2008), pp. 49–71.

Rajalakshmi, U. B., *Udupi Cuisine* (Bangalore: Prism, 2000).

Ram-Prasad, Chakravarthi, "India's Middle Class Failure", *Prospect*, 30 September 2007.

Ramanujan, Attipat Krishnaswami, "Food for Thought: Toward an Anthology of Food Images", in *The Eternal Food: Gastronomic Ideas and Experiences of Hindus and Buddhists*, ed. by Ravindra S. Khare (Albany: State University of New York Press, 1992), pp. 221–50.

Ranade, Mahadeo Govind, *The Rise of Maratha Power* (Bombay: Bombay University Press, 1961).

Ranade, Sanjay, "The Kolis of Mumbai at Crossroads: Religion, Business and Urbanisation in Cosmopolitan Bombay Today", paper presented at the 17th Biennial Conference of the Asian Studies Association of Australia, Monash University, Melbourne, 1–3 July 2008, available at http://artsonline.monash.edu.au/mai/files/2012/07/sanjayranade.pdf [accessed 20 July 2012].

Rane, Kavita, *An Observational Study of Communication Skills Involving Fish Retailers in Mumbai* (unpublished MA thesis, University of Mumbai, 2005).

Rao, Madhugiri Saroja A., "Conservation and Change in Food Habits among Migrants in India: A Study in Gastro-dynamics", in *Aspects in South Asian Food System: Food, Society and Culture*, ed. by Ravindra S. Khare and M. S. A. Rao (Durham, NC: Carolina Academic Press, 1985), pp. 121–40.

Remotti, Francesco, "Cultura", in *Enciclopedia delle Scienze Sociali*, vol. 2 (Rome: Istituto Enciclopedico Italiano, 1992), pp. 641–60.

—, *Contro l'identità* (Rome: Laterza, 1996).

—, *Cultura. Dalla complessità all'impoverimento* (Rome: Laterza, 2011).

Rivera, Annamaria, "Cultura", in *L'imbroglio etnico in quattordici parole-chiave*, ed. by René Gallissot, Mondher Kilani and Annamaria Rivera (Bari: Dedalo, 2001), pp. 75–106.

—, "Neorazzismo", in *L'imbroglio etnico in quattordici parole-chiave*, ed. by René Gallissot, Mondher Kilani and Annamaria Rivera (Bari: Dedalo, 2001), pp. 279–309.

Rotilio, Giuseppe, "L'alimentazione degli ominidi fino alla rivoluzione agropastorale del neolitico", in *In carne e ossa. DNA, cibo e culture dell'uomo preistorico*, ed. by Gianfranco Biondi, Fabio Martini, Olga Rickards and Giuseppe Rotilio (Bari: Laterza, 2006), pp. 83–145.

Roy, Arundhati, "War Is Peace", *Outlook India*, 29 October 2001, available at http://www.outlookindia.com/article.aspx?213547 [accessed 17 July 2012].

— and David Barsamian, *The Checkbook and the Cruise Missile: Conversations with Arundhati Roy* (Cambridge, MA: South End Press, 2004).

Rushdie, Salman, *Imaginary Homelands: Essays and Criticism 1981–1991* (New York: Viking, 1991).

Russell, Robert Vane, *The Tribes and Castes of the Central Provinces of India*, 4 vols. (London: Macmillan, 1916; repr. New Delhi: Asian Educational Services, 1993).

Russell, Sharman Apt, *Hunger: An Unnatural History* (New York: Perseus, 2006).

Sabharwal, Gopa, *Ethnicity and Class: Social Divisions in an Indian City* (New Delhi: Oxford University Press, 2006).

Salsano, Alfredo, *Il dono nel mondo dell'utile* (Turin: Bollati Boringhieri, 2008).

Sapelli, Giulio, "Mitobiografia per le scienze sociali", in *Giannino Bassetti: L'imprenditore raccontato*, ed. by R. Garruccio and G. Maifreda (Soveria Mannelli: Rubbettino, 2004).

—, *Antropologia della globalizzazione* (Milan: Bruno Mondadori, 2002).

Sapir, Edward, *Culture, Language and Personality: Selected Essays* (Berkeley: University of California Press, 1949).

Sardar, Ziauddin and Borin Van Loon, *Introducing Cultural Studies* (Cambridge: Icon Books, 1999).

Sardesai, Govind Sakharam, *New History of Marathas* (Bombay: Bombay University Press, 1946).

Sarkar, Haribishnu and B. M. Pande, *Symbols and Graphic Representations in Indian Inscriptions* (Delhi: Aryan, 1999).

Sassen, Saskia, *Cities in a World Economy* (Thousand Oaks, CA: Pine Forge Press, 1994).

—, *The Global City: New York, London, Tokyo* (Princeton: Princeton University Press, 1991).

Sayad, Abdelmalek, *L'immigration ou le paradoxe de l'altèrité* (Brussels: De Boeck Université, 1991).

Schein, Edgar H., *The Corporate Culture Survival Guide* (San Francisco: Jossey-Bass, 1999).

Segbers, Klaus (ed.), *The Making of Global City Regions: Johannesburg, Mumbai/Bombay, Sao Paulo and Shanghai* (Baltimore, MD: Johns Hopkins University Press, 2007).

Senge, Peter M., *The Fifth Discipline: The Art & Practice of Learning Organization* (Danvers, MA: Doubleday, 1990).

Seppilli, Tullio, "Per un'antropologia dell'alimentazione. Determinazioni, funzioni e significati psico-culturali della risposta sociale a un bisogno biologico", *La ricerca folklorica*, 30 (October 1994), 8–9.

Serventi, Silvano and Françoise Sabban, *La pasta. Storia e cultura di un cibo universale* (Rome: Laterza, 2000).

Shapiro Anjaria, Jonathan, "Street Hawkers and Public Space in Mumbai", *Economic and Political Weekly*, 27 May 2006, pp. 2140–46.

Shiva, Vandana, *Earth Democracy: Justice, Sustainability, and Peace* (Cambridge, MA: South End Press, 2005).

—, *India Divided: Diversity and Democracy Under Attack* (New York: Seven Stories Press, 2005).

—, *Water Wars: Privatization, Pollution, and Profit* (Cambridge, MA: South End Press, 2004).

Sibilla, Paolo, *La sostanza e la forma. Introduzione all'antropologia economica* (Turin: Utet, 1996).

Singh, Rashmi Uday, *Times Food Guide Mumbai 2007* (New Delhi: The Times of India, 2007).

Smith, Brian K., "Eaters, Food and Social Hierarchy in Ancient India: A Dietary Guide to a Revolution of Values", *Journal of the American Academy of Religion*, 58 (1990), 177–206.

—, *Classifying the Universe: The Ancient India Varna System and the Origins of Caste* (New York: Oxford University Press, 1949).

Smith, David, *Hinduism and Modernity* (Oxford: Blackwell, 2003).

Solinas, Pier Giorgio, "Soggetti estesi e relazioni multiple. Questioni di antropologia indianista", *Società Degli Individui*, 25 (2006), available at http://www.antropologica.unisi.it/images/a/ad/L'in-dividuo.pdf [accessed 29 June 2012]. DOI: 10.1400/65102

Sorcinelli, Paolo, *Gli italiani e il cibo. Dalla polenta ai cracker* (Milan: Bruno Mondadori, 1999).

Steel, Carolyn, *Hungry City: How Food Shapes our Lives* (London: Chatto&Windus, 2008).

Stoller, Paul, *The Taste of Ethnographic Things: The Senses in Anthropology* (Philadelphia: University of Pennsylvania Press, 1992).

Taguieff, Pierre-André, *La Force du prejudge. Essai sur le racisme et ses doubles* (Paris: La Découverte; Armillaire, 1988).

Taher, Nasreen, *Impeccable Logistics and Supply Chain Management: A Case of Mumbai Dabbawallahs* (Hyderabad: Icfai University Press, 2007).

Tedlock, Dennis, *The Spoken Word and the Work of Interpretation* (Philadelphia: University of Pennsylvania Press, 1983), it. trans., *Verba Manent. L'interpretazione del parlato* (Naples: L'Ancora del Mediterraneo, 2002).

Thapar, Romila, "Syndicated Moksha?", *Seminar*, 313 (1985), 14–22.

Tindall, Gillian, *City of Gold: The Biography of Bombay* (New Delhi: Penguin, 1982).

Tinker, Irene, *Street Foods: Urban Food and Employment in Developing Countries* (New York: Oxford University Press, 1997).

Torri, Michelguglielmo, *Storia dell'India* (Bari: Laterza, 2000).

Van Binsbergen, Wim M. J., "Cultures Do Not Exist: Exploding Self-Evidences in the Investigation of Interculturality", *Quest: An African Journal of Philosophy*, 13 (1999), 37–144.

Van Gennep, Arnold, *Les Rites de Passage* (Paris: Picard, 1909).

Van Wersch, Herbert W. M., *The Bombay Textile Strike, 1982–1983* (New Delhi: Oxford University Press, 1992).

Varma, Pavan K., *The Great Indian Middle Class* (New Delhi: Penguin, 1998).

Varma, Rashmi, "Provincializing the Global City: From Bombay to Mumbai", *Social Text*, 22:4 (2004), 65–89.

Vicziany, Marika and Jayant Bapat, "Mumbadevi and the Other Mother Goddesses in Mumbai", *Modern Asian Studies*, 43 (2009), 511–41. DOI: 10.1017/S0026749X0700340X

Visconti, Luca, "Diversity management e lavoro straniero: vantaggio competitivo o cerimonia?", in *Diversity management e società multiculturale. Teorie e prassi*, ed. by Luigi Mauri and Luca Visconti (Milan: Franco Angeli, 2004), pp. 11–30.

Von Braun, Joachim, *The World Food Situation: New Driving Forces and Acquired Actions* (Washington, DC: International Food Policy Research Institute, 2007), available at http://www.ifad.org/events/lectures/ifpri/pr18.pdf [accessed 19 July 2012].

Wagner, Roy, *The Invention of Culture* (Chicago, IL: University of Chicago Press, 1981).

Walker Bynum, Caroline, *Holy Feast and Holy Fast: The Religious Significance of Food to Medieval Women* (Berkeley: University of California Press, 1987).

Watson, James L. and Melissa L. Caldwell (eds.), *The Cultural Politics of Food and Eating: A Reader* (London: Blackwell, 2005).

Weber, Max, *The Sociology of Religion*, rev. ed. (London: Methuen, 1965).

West Rudner, David, *Caste and Capitalism in Colonial India: The Nattukottai Chettiars* (Berkeley: University of California Press, 1994).

Wilk, Richard, *Home Cooking in the Global Village: Caribbean Food from Buccaneers to Ecotourists* (Oxford: Berg, 2006).

— and Lisa Cliggett, *Economies and Cultures: Foundations of Economic Anthropology* (Boulder, CO: Westview Press, 1996).

Wittgenstein, Ludwig, "Bemerkungen über Frazers The Golden Bough", *Synthese*, 17 (1967), 233–53.

Wolpert, Stanley, *A New History of India* (Oxford: Oxford University Press, 1977).

Wujastyk, Dominik (ed.), *The Roots of Ayurveda: Selections from Sanskrit Medical Writings* (London: Penguin, 2003).

Yagnik, Acyuta and Suchitra Sheth, *The Shaping of Modern Gujarat: Plurality, Hindutva, and Beyond* (New Delhi: Penguin, 2005).

Zachariah, Kunniparampil Curien, *Migrants in Greater Bombay* (London: Asia Publishing House, 1968).

Zangwill, Israel, *The Melting Pot* (Charleston: Biblio Bazaar, 2008).

Zenner, Walter, "Middleman Minority Theories: A Critical Review", in *Sourcebook on the New Immigration*, ed. by Roy S. Bryce-Laporte, Delores M. Mortimer and Stephen Robert Couch (New Brunswick, NJ: Transaction Books, 1980), pp. 413–25.

Index

This book does not end here...

At Open Book Publishers, we are changing the nature of the traditional academic book. The title you have just read will not be left on a library shelf, but will be accessed online by hundreds of readers each month across the globe. We make all our books free to read online so that students, researchers and members of the public who can't afford a printed edition can still have access to the same ideas as you.

Our digital publishing model also allows us to produce online supplementary material, including extra chapters, reviews, links and other digital resources. Find *Feeding the City* on our website to access its online extras. Please check this page regularly for ongoing updates, and join the conversation by leaving your own comments:

http://www.openbookpublishers.com/isbn/9781909254008

If you enjoyed this book, and feel that research like this should be available to all readers, regardless of their income, please think about donating to us. Our company is run entirely by academics, and our publishing decisions are based on intellectual merit and public value rather than on commercial viability. We do not operate for profit and all donations, as with all other revenue we generate, will be used to finance new Open Access publications.

For further information about what we do, how to donate to OBP, additional digital material related to our titles or to order our books, please visit our website: www.openbookpublishers.com

OpenBook Publishers

Knowledge is for sharing

www.ingramcontent.com/pod-product-compliance
Lightning Source LLC
Chambersburg PA
CBHW071641280326
41928CB00068B/2152